Met. O. 933

METEOROLOGICAL OFFICE

Observer's Handbook

Fourth edition

LONDON

HER MAJESTY'S STATIONERY OFFICE

1982

UDC 551.5(02) : 551.501.1

HMSO publications are available from:

HMSO Publications Centre
(Mail and telephone orders only)
PO Box 276, London, SW8 5DT
Telephone orders 01-873 9090
General enquiries 01-873 0011
(queuing system in operation for both numbers)

HMSO Bookshops
49 High Holborn, London, WC1V 6HB   01-873 0011 (Counter service only)
258 Broad Street, Birmingham, B1 2HE   021-643 3740
Southey House, 33 Wine Street, Bristol, BS1 2BQ   (0272) 264306
9–21 Princess Street, Manchester, M60 8AS   061-834 7201
80 Chichester Street, Belfast, BT1 4JY   (0232) 238451
71 Lothian Road, Edinburgh, EH3 9AZ   031-228 4181

HMSO's Accredited Agents
(see Yellow Pages)

And through good booksellers

ISBN 0 400329 7

CONTENTS

Page

INTRODUCTION.. 1

CHAPTER 1

OBSERVATIONAL ROUTINE

1.1.   Procedure at a full reporting station 3
1.2.   Procedure at stations making abbreviated reports 6
1.3.   Procedure at stations making special reports for aviation 6
1.4.   Procedure at stations making plain-language weather reports 7
1.5.   Procedure at climatological stations 7
1.6.   General remarks on observing 8
1.7.   Time standards and punctuality 9
1.8.   Avoidance of errors 13
1.9.   Observing at night 14
1.10. Care of instruments and equipment 14
1.11. Tabulation and custody of autographic records 17

CHAPTER 2

CLOUDS

2.1.   Historical note 18
2.2.   Observation of clouds 19
2.3.   Classification of clouds 21
2.4.   Cloud amount 29
2.5.   Height of cloud base 31

CHAPTER 3

VISIBILITY

3.1.   Introduction 43
3.2.   Visibility objects 44
3.3.   Determination of visibility during daylight 46
3.4.   Determination of visibility at night 47
3.5.   Vertical visibility 55
3.6.   Runway visual range 56

CHAPTER 4

WEATHER

4.1.   Nature of the observations 57
4.2.   Descriptions of phenomena 58
4.3.   Present weather 67
4.4.   Past weather 72
4.5.   Beaufort letters and international symbols 72

CHAPTER 5

WIND

5.1.   General 80
5.2.   Exposure of surface wind sensors 82
5.3.   Estimation of wind direction and speed 84
5.4.   Wind speed measurement 88
5.5.   Wind direction measurement 91
5.6.   Anemographs 91
5.7.   The Meteorological Office Mk 5B wind system 94
5.8.   Other wind systems 94

Chapter 6

STATE OF GROUND AND CONCRETE SLAB

		Page
6.1.	General	95
6.2.	State-of-ground scales	96
6.3.	State of concrete slab	97

Chapter 7

ATMOSPHERIC PRESSURE

7.1.	General	98
7.2.	Precision aneroid barometers	99
7.3.	Mercury barometers	102
7.4.	Checking barometer readings	105
7.5.	Pressure at aerodrome level (QFE)	105
7.6.	Altimeter setting (QNH)	106
7.7.	Barographs	106

Chapter 8

TEMPERATURE AND HUMIDITY

8.1.	Temperature scales	110
8.2.	Thermometry	110
8.3.	Thermometer screens	117
8.4.	Standard thermometers in the screen	118
8.5.	Thermometers exposed outside screens	125
8.6.	Autographic instruments	129
8.7.	Aspirated psychrometers	132

Chapter 9

PRECIPITATION

9.1.	General requirements	135
9.2.	Basic methods of measurement	135
9.3.	Measuring precipitation by collection	137
9.4.	Measurement of solid precipitation	142
9.5.	Recording rain-gauges	144
9.6.	Gravimetric rain-gauge	151
9.7.	Measurement of snow depth	152

Chapter 10

SUNSHINE

10.1.	General	153
10.2.	Sunshine cards	154
10.3.	Changing the card	154
10.4.	Measuring the trace	156
10.5.	Writing up the cards	158
10.6.	Care of the recorder	158

Chapter 11

SPECIAL PHENOMENA

11.1.	General	160
11.2.	Photometeors	160
11.3.	Special clouds	168
11.4.	Electrometeors	170
11.5.	Miscellaneous phenomena	173
11.6.	Recording phenomena	175

APPENDICES

		Page
I.	Requirements at observing stations	177
II.	Recording of observations for climatological purposes at observing stations	200
III.	Tables	204

BIBLIOGRAPHY 213

INDEX 214

LIST OF DIAGRAMS

Figure		*Page*
1.	Principle of cloud-height measurement with searchlight	34
2.	Measurement of cloud base with searchlight	36
3.	Meteorological Office cloud-base recorder system	42
4(a).	Twilight ⎫	53
4(b).	Moonlight ⎬ nomograms of visibility of lights at night	54
4(c).	Darkness ⎭	55
5.	Sample traces on a tilting-siphon rain-recorder chart	69
6.	Specification of wind direction	81
7.	Schematic drawing of precision aneroid barometer Mk 2	100
8.	Kew-pattern barometer Mk 2	104
9.	Standard types of thermometers	111, 112
10.	Estimating tenths of a degree	115
11.	Wet-bulb coverings	120
12.	Approved types of rain-gauges	136
13.	Reading the rain measure	139
14.	Meteorological Office tilting-siphon rain recorder Mk 2	147
15(a).	Tipping-bucket rain-gauge Mk 3	149
15(b).	Tipping-bucket rain-gauge Mk 5	150
16.	Cross-section of a sunshine recorder bowl	155
17.	Diagram showing position of sunshine trace at different times of the year when the latitude setting of the instrument is correct	159
18.	Halo phenomena	161
19.	Auroral forms	172
20.	Layout of an observing station	179
21.	Isogonals or lines of equal westerly declination, 1977·5	182
22.	Turf wall for use at exposed rain-gauge sites	189
23.	Variations of the sun's altitude and azimuth	191–194

LIST OF PLATES

Plate   *To face page*

I. { Cirrus fibratus ....
   { Cirrus uncinus ....

II. { Cirrus intortus ....
    { Cirrocumulus lacunosus ....

III. { Cirrocumulus undulatus ....
     { Cirrostratus with cirrus fibratus ....

IV. { Cirrostratus nebulosus with halo ....
    { Altocumulus stratiformis, translucidus, perlucidus ....

V. { Altocumulus lenticularis ....
   { Altocumulus floccus ....

VI. { Altocumulus castellanus ....   between
    { Altocumulus and altostratus opacus, supplementary feature   pages 26
          mamma ....   and 27

VII. { Altostratus translucidus ....
     { Altostratus opacus ....

VIII. { Nimbostratus with stratus fractus ....
      { Stratus fractus ....

IX. { Stratus nebulosus, opacus ....
    { Stratocumulus stratiformis, perlucidus ....

X. { Stratocumulus stratiformis, undulatus ....
   { Cumulus humilis ....

XI. { Cumulus congestus ....
    { Cumulus mediocris ....

XII. { Cumulonimbus congestus with cumulonimbus calvus ....
     { Cumulonimbus capillatus ....

XIII. { Meteorological Office cloud searchlight ....
      { Meteorological Office alidade ....

XIV. Cloud-base recorder charts ....
XV.   Meteorological Office visibility meter Mk 2 ....

XVI. { Soft rime ....
     { Glaze ....

XVII.   Meteorological Office hand anemometer ....
XVIII. Meteorological Office electrical anemograph Mk 4 ....
XIX.   Electrical anemograph record ....

XX. { Meter display for the Mk 5 wind system ....
    { Meteorological Office anemograph recorder Mk 5 ....

XXI.   Precision aneroid barometers Mk 1 and Mk 2 ....

XXII. { Static pressure head ....
      { Open-scale barograph ....

XXIII. Temperature indicator Mk 5 ....

XXIV. { Digital temperature indicator Mk 1A..   between
      { Arrangements of instruments in a large thermometer screen ....   pages 106

XXV.   Large thermometer screen...   and 107

XXVI. { Bimetallic thermograph ....
      { Hair hygrograph ....

XXVII.   Aspirated psychrometer ....
XXVIII. Meteorological Office tilting-siphon rain recorder Mk 1 ....
XXIX.   Sunshine recorder Mk 2 ....
XXX.   Sunshine recorders Mk 3C ....
XXXI.   Measurement of sunshine cards ....

Plates I–V (upper), VII–X, XI (lower) and XII are by courtesy of Mr R. K. Pilsbury; Plates V (lower), VI (upper), XI (upper) and XVI are by courtesy of Mr S. D. Burt; Plate VI (lower) is by courtesy of Mr C. S. Broomfield; Plates XIII, XXV and XXX (lower) are by courtesy of the Photographic Section, RAF Benson; Plates XXIV (lower), XXVI and XXVIII are by courtesy of the Photographic Section, RAF Leuchars.

OBSERVER'S HANDBOOK

INTRODUCTION

Meteorological observations are made for a variety of reasons. Those observations made primarily for the purpose of providing information for weather forecasts are termed 'synoptic'. These synoptic observations are not by themselves sufficient to meet all the needs of, for example, agricultural meteorology, hydrology and industry. Hence, in the United Kingdom there is a further network of voluntary co-operating stations maintained by private individuals, schools and colleges, industrial concerns, local authorities, etc. whose records supplement those from the synoptic reporting stations. In addition there is a much larger number of 'rainfall stations' where the only records regularly maintained are those of rainfall.

The aim of all these observations is essentially to provide data. The differences between them arise mainly from the use to which the data are initially put. Synoptic observations are more frequent and more detailed, and are encoded for immediate transmission to forecasting centres. The other stations report in a manner suited to their particular function and ability, and are detailed below. All types of station should aim at high standards of accuracy, punctuality and instrument care. From this generalized classification there emerge five categories of station in the United Kingdom to which this handbook is addressed:

(a) Synoptic stations whose primary function is to provide data for the forecasting service and for aviation purposes.

(b) Auxiliary reporting stations also make observations which supplement the main network of synoptic stations. The number of observations made, and the type, vary according to a particular station's circumstances. The contributions may range from an observation every hour to one every three hours for a limited period.

(c) Climatological stations are manned by voluntary co-operating observers who make a daily observation at 0900 GMT. These observations are not reported at the time but they make a significant contribution to the long-term climatological data for the British Isles.

(d) Health Resort stations at which, in addition to the normal procedure of a climatological station, the observers make special reports at 6 p.m. clock time for inclusion in bulletins to the Press by the Meteorological Office.

(e) Agricultural meteorological (agrometeorological) stations which are maintained for research and operational decisions on field-work (mostly by the Ministry of Agriculture, Fisheries and Food). These stations also make an 0900 GMT climatological observation.

The majority of the stations described above make an additional contribution in the form of climatological summaries which are completed monthly and, in some cases in an abbreviated form, weekly. The Meteorological

Office maintains a number of stations undertaking specialized functions which are outside the scope of this handbook; these include the Observatories and upper-air stations. The former have a multi-purpose role, making additional reports such as solar radiation, atmospheric pollution, evaporation, atmospheric electricity, etc. The upper-air stations make measurements of temperature, humidity and wind in the upper air, and some of these stations are involved in the location of thunderstorms, in solar radiation measurements and in surface observations.

This handbook has been based largely on guidance provided by the World Meteorological Organization (WMO). In particular, much use has been made of the WMO *Guide to meteorological instrument and observing practices,* 1971, and the WMO *International cloud atlas,* Volume I, 1975.

It is assumed throughout this book that the observer is at a station which has been properly sited and equipped, on the scale appropriate to its type, and that the station is in full working order. Chapter 1 accordingly begins with a brief summary of the observing procedure at each type of station and gives some general notes on observing. The chapters which follow are devoted to the details of each type of observation. Reference is made to the use and maintenance of instruments where necessary in describing observational techniques. Since the previous (3rd) edition of this handbook was published some new instruments have been brought into routine use. These are described in general terms, but for more complete advice on instruments the reader is referred to the *Handbook of meteorological instruments* and either to the special maintenance instructions or the installation/operator instructions which are published by the Operational Instrumentation Branch of the Meteorological Office. The main text is followed by three Appendices, one of which gives directions for the selection of the site and for those other matters which must receive attention when the station is being set up; thus it is only occasionally necessary to refer to questions of exposure in the body of this book.

The metric (SI) system of units is becoming increasingly accepted in this country and is used throughout this edition. For a time Imperial units are likely to remain in use for some purposes, notably for reporting wind speed and cloud height. In these cases the Imperial units are added where these are considered important and informative.

Except where otherwise indicated, all times refer to Greenwich Mean Time (GMT) and are specified in the 24-hour system beginning at 0000. To emphasize the distinction between GMT and time by public clocks, the a.m. and p.m. system is used when reference is made to clock time. Adequate explanations accompany the few references made to other time standards, namely Local Zone Time and Local Apparent Time.

Details of the books mentioned throughout this handbook are given in the Bibliography on page 213.

Observers who still have a requirement for correction tables applicable to mercury barometers will need to retain the previous (3rd) edition of this handbook (see the introductory text in Appendix III, page 204).

CHAPTER 1

OBSERVATIONAL ROUTINE

1.1. PROCEDURE AT A FULL SYNOPTIC STATION

1.1.1. Observations at each scheduled hour are usually made in the following order:

Temperature, dry-bulb
Temperature, wet-bulb (used in conjunction with the dry-bulb reading to derive the vapour pressure, relative humidity and dew-point)
Amount of precipitation (selected stations only)
Amount of cloud
Types of cloud
Height of cloud base
Present weather
Visibility
Wind direction and speed
Pressure tendency and characteristic
Atmospheric pressure
Past weather and any special phenomena occurring either at the time of observation or since the last observation are also recorded.

The main synoptic hours of observation are 0000, 0600, 1200 and 1800 GMT. Intermediate synoptic hours are 0300, 0900, 1500 and 2100 GMT.

1.1.2. Additional routine observations made in the United Kingdom at certain specified hours. These include:

Extreme screen temperatures (maximum and minimum): read and reset at each of the hours 0600, 0900, 1800 and 2100 GMT
Grass minimum temperature: at 0900 GMT (also at 0600 GMT without resetting)
Soil temperatures: at 0900 GMT at selected stations
Concrete-slab minimum temperature: at 0900 GMT
State of concrete slab: at 0900 GMT
State of ground: 3-hourly at 0000, 0300, 0600, . . . 2100 GMT
Amount of precipitation: a distinction is made here between rainfall measurements made for climatological purposes and those made for synoptic purposes. Climatological measurements are made at 0900 and 2100 GMT (or 0900 only). Synoptic stations will, in addition, measure rainfall at the main synoptic hours, reporting at 0600 and 1800 GMT the previous 12-hour total rainfall, and at 0000 and 1200 GMT the previous 6-hour total rainfall. Selected stations also report the hourly rainfall. Measurements of rainfall for synoptic purposes will be made using another, quite separate, gauge for the purpose. A separate gauge is used to preserve the integrity of the measurements made for climatological purposes.

Snow depth: the depth of accumulated undrifted snow is reported (see the *Handbook of weather messages,* Parts II and III) by synoptic stations at 0600 GMT. All stations making synoptic observations will also report the snow depth at 0600, 0900, 1800 and 2100 GMT whenever snow is lying to an undrifted depth of 0·5 cm or more and additionally, as specified in 9.7 on page 152, at any other main or intermediate synoptic hour.

Note that these measurements are separate from the measurement of the water equivalent which is reported under rainfall; the water equivalent is dealt with in 9.4 on page 142.

Duration of bright sunshine: at stations equipped with sunshine recorders the card is changed daily as soon as possible after sunset. Exceptionally, the cards may be dealt with in accordance with 10.3 and 10.5, pages 154 and 158 respectively.

1.1.3. Selected stations include the following supplementary data in a separate message at 0900 GMT and 2100 GMT.

At 0900 GMT:

Maximum temperature ⎫ for the period 2100 GMT (previous day) to
Minimum temperature ⎬ 0900 GMT, or 0900 GMT (previous day)
Amount of precipitation ⎭ to 0900 GMT if the station does not report
at 2100 GMT

Grass minimum temperature ⎫
Soil temperature at 30 cm and 100 cm ⎬ read at 0900 GMT
Concrete-slab minimum temperature ⎭
State of concrete slab at 0900 GMT
Depth of snow at 0900 GMT
Depth of fresh snow at 0900 GMT
Total hours of bright sunshine for previous day
Occurrence of hail, thunder, gale, snow or sleet on previous day
Existence of fog at 0900 GMT on previous day
States of ground three-hourly from 1200 GMT (previous day) to 0900 GMT.

At 2100 GMT:

Maximum temperature ⎫
Minimum temperature ⎬ for the period 0900 GMT to 2100 GMT.
Amount of precipitation ⎭

1.1.4. Autographic instruments.

The charts on most autographic instruments are changed either daily or weekly. Some, for example the electrical anemograph, record on roll charts lasting for one month. The charts on instruments with daily clocks are changed at the time of the 0900 GMT observation, those with weekly clocks are changed about 0900 GMT on Mondays; monthly roll charts are normally changed at 0000 GMT on the first day of the month. If this day falls on a Saturday or Sunday at stations which are not manned at the weekend, the change is made on the first Monday of the month.

Before a chart is changed, the name and details of the station, dates and serial number of the chart are entered in ink on the new chart. A time mark is made one or two hours after a chart has been changed, and on weekly and monthly charts thereafter at about 1200 GMT daily (or 1800 GMT if more convenient). On daily charts it is useful to have a second time mark 8–12 hours after the first. All time marks are recorded to the nearest minute; an appropriate note is entered in the Register and details of date and time are transferred to the chart after removal. For anemograms it is preferable to annotate the chart at the time the mark is made.

The appropriate measured value and the time are noted in pencil on each chart at the beginning and end of each trace. After each chart has been taken off the instrument, these and all other details required on the chart are completed neatly in ink. On daily thermograms and hygrograms, values of dry-bulb temperature (screen) and relative humidity (measured by psychrometer) are noted on the respective charts at the time of observation at the main synoptic hours, 0000, 0600, 1200 and 1800 GMT, whenever possible. On weekly barograms the mean-sea-level (MSL) pressure, measured by precision aneroid barometer or mercury barometer, is noted against the daily time marks.

The serial number for each chart is obtained by reference to Met. O. Leaflet No. 11, The Meteorological Office Calendar, published annually.

Where autographic instruments are maintained in a screen (thermograph and hygrograph) or in the open (recording rain-gauge), charts may be exposed to precipitation during changing. It is advisable to provide them with some form of protection both before and after exchanging charts; a waterproof folder or a suitable container can help to avoid unnecessary spotting of traces and excessive dampness of the chart. It is unavoidable at times and, if it occurs, charts must be dried before attempting to enter the chart details in ink.

All autographic instruments must be inspected from time to time during the day to make sure they are working correctly. Care must be taken that the observer's daily familiarity with the instruments does not lead him to overlook the deterioration that takes place in instrument performance over a period of time. Gradually accumulating deposits on nibs, clock timing incorrect, and a slow shift in instrument calibration are examples of deteriorations to which an observer can grow accustomed but which can be corrected. See also 1.10 and 1.11, pages 14–17.

1.1.5. Pilot-balloon observations are made at Meteorological Office stations as necessary to meet special requirements. Instructions are given in *Measurement of upper winds by means of pilot balloons.*

1.1.6. Recording observations. Observations (other than pilot-balloon observations) are recorded in the Daily Register (Metform 2050). Detailed instructions for making the entries are given in the *Handbook of weather messages,* Part III. A pocket-size pad of observation slips (Metform 2051) is provided for use when the observations are being made, and the recorded observations are copied from the slip into the Daily Register. Care must be taken at this stage to ensure that the transcription of the entries from the observation slip to the Register is correct.

1.2. PROCEDURE AT STATIONS MAKING ABBREVIATED REPORTS

Many auxiliary reporting stations make abbreviated reports in a code, the details of which are given in the *Handbook of weather messages*, Part III, and also in the publication *Abbreviated weather reports*. The latter also contains details for the entry of all such observations in Metform 2611 (Register of observations). The hours for observations are not the same for all stations but are arranged individually to suit local requirements or to suit the times when the observer is available. The following observations are required at each hour of observation:

> Temperature, screen dry-bulb } if required by Meteorological
> Dew-point   } Office Headquarters
> Amount of cloud
> Types of cloud
> Heights of cloud
> Present weather
> Visibility
> Wind direction and speed
> Past weather

Some stations report supplementary data at 0900 and 2100 GMT (or at 0900 GMT only). They report these data in the form in the *Handbook of weather messages,* Part III, and in *Abbreviated weather reports.*

1.3. PROCEDURE AT STATIONS MAKING SPECIAL REPORTS FOR AVIATION

1.3.1. Special reports.   To meet the needs of aviation, meteorological offices situated on aerodromes are required to make additional reports. The times, frequency and format of these additional reports are the subject of special instructions to the stations concerned. The two most common forms of report are outlined below.

1.3.1.1. *Reports of sudden changes.*   These reports are identified by the preliminary letters MMMMM, indicating a deterioration, or BBBBB, indicating an improvement, in certain elements either below or above certain defined limits; (see the *Handbook of weather messages,* Part III, for the conditions governing the making of these reports). The weather is to be kept constantly under observation to ensure that occasions when changes in weather conditions demanding these reports are not missed.

1.3.1.2. *Aviation routine weather reports.*   An abbreviated report (METAR) in a form specifically for use for aviation purposes is given in the *Handbook of weather messages,* Parts II and III. The content of this report, which can also contain plain-language elements, includes:

> Wind direction and speed
> Visibility
> Runway visual range (if required)
> Present weather

Amount and types of cloud
Height of cloud base
Screen dry-bulb temperature }
Dew-point   } (if required)
Pressure: QNH value (see 7.6, page 106).

1.4. PROCEDURE AT STATIONS MAKING PLAIN-LANGUAGE WEATHER REPORTS

Simplified weather reports, in plain language, are made at a number of Auto-mobile Association and Royal Automobile Club town offices and Motorway Maintenance Depots. These reports, which are entered on Metform 5967B or D (Plain-language weather observations), indicate the state of sky, the weather, the visibility and state of roads. Some stations also report dry-bulb and minimum temperatures. The schedule for these reports varies; usually only two or three reports a day are made from any one site, but in winter, when motorway maintenance compounds are manned continually, 3-hourly reports throughout the 24 hours are made at many of these sites.

Reports are telephoned to a collecting centre and are used for forecasting in the public service sector.

1.5. PROCEDURE AT CLIMATOLOGICAL STATIONS

1.5.1. Climatological stations, manned by voluntary co-operating observers, normally make the following instrumental and visual observations once a day at 0900 GMT:

Temperature, dry-bulb, in the screen
Temperature, wet-bulb, in the screen
Temperature extremes, maximum and minimum, in the screen (both thermometers are then reset)
Amount of precipitation 0900–0900 GMT (rain in the measuring glass is then poured away)
State of ground (see Chapter 6)
Snow depth (see 9.7, page 152)
Amount of cloud
Present weather
Visibility
Past weather: a brief account for the previous day, midnight to midnight GMT, is kept and entered on the monthly return.

The above observations are the basic requirements from a climatological station but those who wish to do so, providing the site is satisfactory, may add instruments to indicate or record the following:

Sunshine duration (see 10.4, page 156)
Temperature, grass minimum (thermometer then reset)
Temperature, bare-soil minimum (thermometer then reset)
Temperature, concrete-slab minimum (thermometer then reset)
Temperature, soil (as may be arranged for individual stations)
Wind: direction in tens of degrees from true north, and speed in knots

Atmospheric pressure (readings from a precision aneroid or mercury barometer).

At all climatological stations the observations are recorded in Metform 3100 (Pocket Register) and then transcribed on to a monthly return. Instructions for completing this return are given in Metform 3100A which is supplied to observers voluntarily co-operating with the Meteorological Office.

1.5.2. Stations at Health Resorts.   A number of coastal and inland resorts participate in the Meteorological Office Health Resort Scheme. In addition to carrying out the 0900 GMT climatological observations listed in 1.5.1, Health Resort stations make an observation at 6 p.m. daily. These reports are entered in Metform 3100 and are encoded for telephoning to a collecting centre. Details of the collecting centres used by individual stations and the form of the coded message are notified to each station.

The report made at 6 p.m. is intended for release to the Press and includes two groups of Beaufort letters (see 4.5.1, page 73), each not exceeding five in number, which describe the past weather from 8 a.m. to noon, and noon to 6 p.m. respectively. Alternatively, if a single word can be used to describe the weather throughout either of those periods, this word can be substituted for the group of Beaufort letters. In addition, the following observations are made and reported at 6 p.m. clock time:

Temperature, dry-bulb, in the screen
Temperature, maximum, in the screen (the thermometer is not reset)
Temperature, minimum, in the screen, as read at 0900 GMT on the day of the report
Amount of precipitation. (The two totals reported are for the periods 6 p.m. on the previous day to 0900 GMT, and 0900 GMT to 6 p.m. on the day of the report. The rain in the measuring glass is poured back into the rain-gauge after the 6 p.m. measurement.)
Sunshine duration. (The two totals reported are for the periods 6 p.m. to sunset on the previous day, and from sunrise to 6 p.m. on the day of the report.)

At Health Resort stations the sunshine cards may be changed at 6 p.m. or between sunset and sunrise as the observer prefers. Whichever practice is adopted, it must be maintained regularly. The special procedure for changing the cards is explained in 10.3, page 154.

1.5.3. Agricultural meteorological stations.   These are climatological stations maintained by those with a direct interest in the effects of weather on agriculture. These stations carry out the routine in 1.5.1 above and, in addition, record at 0900 GMT the soil temperatures at one or more depths, bare-soil minimum temperature, and usually the run of wind at a height of 2 metres. Some stations also record solar radiation.

1.6. GENERAL REMARKS ON OBSERVING

The preceding paragraphs merely list the items to be observed. Details of the items to be observed in each case are set down in the chapters which follow.

Sufficient information is given here for observers at climatological stations to complete the 0900 GMT daily observations without reference to other publications, but Metform 3100A is required for instructions for completing the monthly returns. Observers at other types of station must refer to publications indicated in the preceding paragraphs and which are repeated below for convenience.

(*a*) At stations which make abbreviated synoptic reports, observers refer to *Abbreviated weather reports.*

(*b*) At full synoptic stations, observers refer to the *Handbook of weather messages,* Parts II and III.

(*c*) At aerodromes where special reports are made for aviation by meteorological staff, observers refer to the *Handbook of weather messages,* Parts II and III. Where reports are made by Air Traffic Control staff, reference is more conveniently made to a wall-card—*Codes for observations in METAR form.*

There is no essential difference between the observation of items at a climatological station and at a synoptic station. At a synoptic station where the schedule of observations is hourly throughout the 24 hours, or covers at least several hours each day, it is essential that the watch on the weather is as close and continuous as possible. At a climatological station where normally only one observation a day is recorded (at 0900 GMT) and where, for example, the observer may be an employee of a local authority, or someone with other duties, or a private person making observations as a matter of interest, the maintenance of a record of the weather occurring between the hours of observation is very much more difficult, but no less important. The observer has to try to make his record as complete and as reliable as is practicable and, if possible, state clearly the period of the day covered by the weather diary. It will provide not only the basis for the preparation of climatological statistics, which are used for a wide variety of scientific and industrial purposes, but it may also have to be produced in a Court of Law.

1.7. TIME STANDARDS AND PUNCTUALITY

In the United Kingdom the standard of time for all meteorological observations is Greenwich Mean Time (GMT). When time by public clocks is one hour ahead of GMT an observation for 0900 GMT, for example, is made at 10 a.m. clock time.

At overseas stations manned by Meteorological Office staff, the standard of time for all synoptic purposes is also GMT, but for climatological purposes the local zone time (LZT) is used (see 1.7.7). This LZT is that appropriate to the meridian of a station, though a variation of up to one hour is permissible where such variation would bring the station's practice into accord with local time standards. The standard of time used at an overseas station should be plainly stated in the Daily Register, on monthly returns, and on all records and tabulations.

1.7.1. Standard hours of observations for synoptic purposes. As decided by the World Meteorological Organization, GMT is standard throughout the world for synoptic purposes. The main standard times for surface synoptic observations are 0000, 0600, 1200 and 1800 GMT; the intermediate standard times are 0300, 0900, 1500 and 2100 GMT. Standard times for surface observations at other than main and intermediate hours are 0100, 0200, 0400, 0500 GMT, etc. The evaluation of elements composing a surface synoptic observation should be made in as short a time as possible, just prior to the official time of observation. About 10 minutes should normally suffice to complete the observation.

1.7.2. Hours of climatological observations at stations in the United Kingdom.   At climatological stations, and at stations which measure rainfall only, the standard hour of observation is 0900 GMT. Health Resort stations are a special case, and for these stations a special evening observation is made at 6 p.m. clock time throughout the year in addition to the climatological observation at 0900 GMT. The reason for this exceptional use of clock time is that special reports have to be sent to newspapers at times related to that of going to press. Synoptic stations may be required to compile climatological returns for a variety of hours as determined by other instructions.

1.7.3. Hours of climatological observations at stations outside the United Kingdom.   At overseas synoptic stations manned by Meteorological Office staff, monthly returns are required for the eight synoptic hours (0000, 0300, . . . 2100 GMT) at which the observations are made. The climatological terminal hours for the measurement of amounts of precipitation and extreme temperatures are 0900 and 2100 local zone time (LZT), or the hours of synoptic observation nearest to 0900 and 2100 LZT.

1.7.4. Hours of observing amount of precipitation and extreme temperature at synoptic stations: European (Region VI) practice.   Rainfall amounts are included by full synoptic stations at 0000, 0600, 1200 and 1800 GMT in their weather reports. The amounts reported are for the previous six hours at 0000 and 1200 GMT, and the totals of the previous twelve hours at 0600 and 1800 GMT. Extremes of temperature (maximum and minimum) are observed at 0600, 0900, 1800 and 2100 GMT. The method of obtaining the values to be included in synoptic reports is explained in the Daily Register.

Additionally in the United Kingdom, as noted in 1.1.2, measurements of rainfall are made at 0900 and 2100 GMT (or at 0900 only) for climatological purposes; at certain stations hourly rainfall reports are made.

1.7.5. Punctuality is essential in all meteorological observing.   The observations of the elements composing a weather report cannot all be made at once, but the World Meteorological Organization has recommended that the pressure observations should be made at the exact hour, that visibility should normally be the last of the outdoor observations to be made, and that the other elements should be observed in the 10 minutes immediately preceding the hour. However, at synoptic stations, timing schedules for onward transmission of the reports to the collecting centre may necessitate the completion of the observation before the exact hour: this may vary from 5 to 10 minutes. When any element changes significantly during the 10-minute period of observation, and the observer is able to assess this change, the new assessment of the changing element may be the last thing to be observed.

Definitions of 'times' of observations are given below in 1.7.5.1 to 1.7.5.3.

The order of the observations advised in 1.1.1, 1.2 and 1.5.1 is usually the most convenient at stations where the layout follows the plan suggested in Appendix I (page 179) and where instruments record or indicate surface wind speed and direction remotely; where there are no such instruments, surface wind speed and direction should be observed before the visibility. The order of precedence has to be varied in special circumstances, but should ensure that all the scheduled elements are observed within the 10 minutes before the official time of observation.

If a balloon observation is made to determine the height of the cloud base, sufficient time must be allowed to ensure that the normal observation of all other elements is completed by the required time.

1.7.5.1. *Actual time of observation* of an element means the time at which the observation of that particular element is completed. For any element which is changing during the period when an observer is making his observation, the actual time of observation of that element should approximate as closely to the 'official time of observation' as will permit the observer to assess the change.

1.7.5.2. *Official time of observation* is the time of observing the last element necessary to complete a surface synoptic observation. This will normally be the reading of the barometer, but when any element changes and the observer is able adequately to assess the change, the new assessment of this element may be the last to be observed. This official time of observation is entered in the Daily Register (and, where applicable, on Metform 2309). This time should be as near as possible to the standard time of observation.

1.7.5.3. *Standard time of observation* means the internationally agreed time as contained in World Meteorological Organization resolutions (see 1.7.1).

1.7.6 Numbering of hours.   When recording the time of occurrence of phenomena etc. the four-figure system is always used. The first two figures give the hour from 00 to 23 and the last two figures give the minutes. For example, 25 minutes past midnight is recorded as 0025, and 25 minutes before midnight as 2335. The day is regarded as beginning at 0000 GMT, and this is logged as the standard time of the first observation of the day, not as the last of the preceding day.

1.7.7. Specification of zone time.   Although GMT is universally adopted as standard for the making of synoptic reports in all parts of the world, it is sometimes necessary to specify time in terms of zone time. In this system the local time varies according to the longitude in steps of an hour per unit of 15 degrees of longitude, and a letter is appended to the four-figure time group to indicate the zone, according to the following scale:

Z = GMT

A, B, C, . . . L, M (omitting the letter J) are zone times appropriate to areas centred on longitudes 15°E, 30°E, 45°E, . . . 165°E, 180°(E).

N, O, P, . . . X, Y are zone times appropriate to areas centred on longitudes 15°W, 30°W, 45°W, . . . 165°W, 180°(W). The complete scheme is given in the table on page 12.

ZONES FOR LONGITUDE EAST OF GREENWICH, GMT BEING BEHIND ZONE TIME

Index letter of time standard	Z	A	B	C	D	E	F	G	H	I	K	L	M
Correction to obtain GMT from zone time (hour)	0	−1	−2	−3	−4	−5	−6	−7	−8	−9	−10	−11	−12
Standard longitude of corresponding zone	0°	15°E	30°E	45°E	60°E	75°E	90°E	105°E	120°E	135°E	150°E	165°E	180°(E)

ZONES FOR LONGITUDE WEST OF GREENWICH, GMT BEING AHEAD OF ZONE TIME

Index letter of time standard	Z	N	O	P	Q	R	S	T	U	V	W	X	Y
Correction to obtain GMT from zone time (hour)	0	+1	+2	+3	+4	+5	+6	+7	+8	+9	+10	+11	+12
Standard longitude of corresponding zone	0°	15°W	30°W	45°W	60°W	75°W	90°W	105°W	120°W	135°W	150°W	165°W	180°(W)

Examples:

$$1300D = 0900Z \qquad 1300M = 0100Z \qquad 1300R = 1800Z$$

$$\left. \begin{matrix} 1400M \\ 0800F \\ 0200Z \end{matrix} \right\} \text{ with a given date, } d \qquad \left. \begin{matrix} 2000S \\ 1400Y \end{matrix} \right\} \text{ with the preceding date, } d - 1$$

The zones for which the time standards have index letters M and Y are separated by the international date-line. Thus 0600M is the same as 0600Y but the latter is dated one day earlier than the date of the former. The following times are identical:

1.7.8. Sunrise and sunset.   The World Meteorological Organization (WMO) recommended in 1957 that, in accordance with current astronomical practice, the definitions of sunrise and sunset should refer to the times when the sun's upper limb contacts the apparent horizon. From July 1964, the Meteorological Office adopted the WMO recommendations: sunrise and sunset are now defined as the instant at which the upper limb of the sun appears in contact with the horizon, it being assumed that both observer and horizon are at sea level. The 'duration of daylight' or 'length of day' are terms used for the interval between sunrise and the following sunset. This quantity may be obtained from the *Nautical almanac* or, more simply, from Table 171 of the *Smithsonian meteorological tables,* pages 506–512 (Sixth revised edition).

The times of sunrise and sunset vary with latitude and with the declination of the sun. Diagrams illustrating the variations are given in Figure 23 on pages 191–194. If these times are required with any accuracy for a station, reference will have to be made to the current edition of the *Nautical almanac* (published annually).

Copies of both the *Nautical almanac* and the *Smithsonian tables* are available for reference in the National Meteorological Library at Bracknell and in most Public Libraries.

1.8. AVOIDANCE OF ERRORS

By approaching his task systematically an experienced observer should be able to complete a set of observations in 10 minutes without sacrificing accuracy to speed. The program must be carried out briskly but not hastily. In the chapters which follow, attention is drawn to the sources of error likely to arise in various types of observation, but it is important to note the need for checking entries immediately after they are made in the Register. This does not mean that all the readings should be repeated. The checking ensures that nothing has been omitted and that no gross error has been made by sheer inadvertence. One of the commonest of such errors is the misreading of a thermometer by 5 or 10 whole degrees; when any instrument is read it is advisable to take a second glance to avoid elementary mistakes of this kind.

Another object in checking the entries is to ensure that they are mutually consistent. If some rain is measured there must be an entry of precipitation somewhere in the 'past weather' column. Fresh snow lying must similarly be accounted for. There should be no inconsistency between entries of visibility and present weather or past weather. Such examples might be multiplied almost indefinitely and the point need not be laboured. At synoptic stations, where observations are initially written on the observations slip (Metform 2051) and then copied into the Daily Register, great care should be taken to avoid errors in copying.

The Register is the official record of all surface observations made. It is therefore the basic document to which reference has to be made when answering enquiries which involve the observations. Corrections must never be made by erasure of the original entry, but are to be made by following the procedures given in section 7.4 of the *Handbook of weather messages,* Part III and in *Abbreviated weather reports.*

1.9. OBSERVING AT NIGHT

Outdoor instrumental observations at night, requiring the use of a light, must be completed at least two minutes before non-instrumental observations to give the observer's eyes time to adapt to the dark (but see the special instructions given in 3.4.1, page 48, concerning the use of the Meteorological Office visibility meter).

1.9.1. Cloud form and amount.   When cloud observations are made at night the principle of continuity is to be borne in mind. The observer should watch the sky towards dusk and so obtain some guidance as to the clouds which are likely to be present after the daylight has gone. If, during the hours of darkness, there is an onset of precipitation and the observer is in doubt as to the form of clouds present, the table on page 29 will assist him in identifying the form of clouds by the character of the precipitation.

If the air is clear, the estimation of total amount of cloud should not usually be difficult because stars will be visible in the areas of the sky free from cloud, but the brightest stars and planets are visible through thin veils of cirrocumulus, cirrus or cirrostratus. The intensity of the darkness is of some assistance in deciding whether the sky is wholly covered or not with dense low cloud; if there is any light at all, variation of contrast or luminance may indicate patches of low cloud with medium or high cloud above.

On occasions of fog which is so thick as to make it impossible to tell whether or not there is cloud above, the amount is recorded as 9 but taken as 8 for statistical purposes, and cloud types are shown as '/'. If cloud can be seen through the fog, the amount of cloud is to be estimated as well as circumstances permit. If the moon or stars can be seen through the fog and there is no evidence of cloud, the cloud amount is to be recorded as 0 (see 2.4.1, page 30).

1.9.2. Cloud height.   The cloud height at night is to be determined whenever possible with the assistance of a cloud-base recorder, cloud searchlight, or by balloon and lantern. Where no instruments are available but there is sufficient light for the type of cloud to be distinguished, the height is to be estimated in the same way as in daylight to the best of the observer's ability (see 2.5.3, page 32).

1.9.3. Visibility.   Instructions relating to determining visibility at night are given in 3.4 (page 47).

1.9.4 Weather.   In observing weather at night the observer must be on his guard against mistaking the flashes created by electric trains for distant lightning.

1.10. CARE OF INSTRUMENTS AND EQUIPMENT

It is essential that the meteorological instruments and equipment on a station be kept clean and, by simple routine maintenance, kept in good working order; this will promote their good performance and lengthen their useful life.

Details of the routine maintenance that must be carried out on a specific instrument will be found in the appropriate chapter of this handbook. A notebook should be maintained in which each instrument is allotted several pages so that a chronological record of routine maintenance, checks and adjustments can be logged. The pages allotted to each instrument can be vertically ruled for appropriate headings; for example, under the section allocated to the thermograph:

Date	Instrument number	Job	Remarks	Initials
5.3.82	345/71	Monthly service	Instrument cleaned, bearings lubricated, temperature setting adjusted	ABC

In addition, a brief note of the action taken and the time it was done should be entered in the remarks column of the observation Register.

1.10.1. General principles. Some general principles to be followed in handling the mechanical parts of recording instruments are given below.

(a) Consult the Installation/Operator's Instructions issued by the Operational Instrumentation Branch of the Meteorological Office.

(b) Do not use unsuitable cleaning materials (e.g. of a corrosive or clogging nature).

(c) Use the correct lubricant, usually good quality clock oil. Apply it sparingly and only to bearings where it is needed, and remove any surplus.

(d) Do not defer cleaning or the routine maintenance until the instrument has reached a condition of neglect which prevents it from working properly. Cleaning should be done regularly, and all instruments should be maintained in good working order.

(e) If a tool needs to be used on an instrument, perhaps for the purposes of adjustment, then use the correct tool of the correct size as this will avoid damage. Never use a knife blade as a substitute for a screwdriver.

(f) Replace any instrument as soon as possible if it appears to be unserviceable. Spares are normally held of those items particularly liable to damage or breakage, e.g. thermometers.

Store all equipment, and spares in particular, so that they are safe from loss or damage but are readily available for use. Ensure that replacements are speedily arranged for unserviceable instruments and for spares which have been used.

1.10.2. Methods of cleaning. General advice and methods of cleaning the different materials most often used in instruments are listed below.

(a) *Plain brass or copper parts.* These may be kept bright with an oily rag or with metal polish applied sparingly. The inside of a rain-gauge funnel, however, should only be rubbed with a damp cloth.

(b) *Lacquered brass or copper parts.* Clean with a soft lint-free cloth. Where there is exposure to damp, a little petroleum jelly may be applied.

(c) *Polished woodwork* (cases of barographs etc.). Clean with a soft lint-free cloth. A little linseed oil may be rubbed in with a soft cloth if necessary.

(*d*) *Glass and porcelain* (thermometer stems, windows of recording instruments).   Clean off dirt with a moist rag or lint-free cloth.

(*e*) *Plain bearings, pinions and hinges of instrument cases.*   Lubricate sparingly with a touch of good quality clock oil (but see 1.10.1(*a*) above).

(*f*) *Steel parts.*   Clean with an oily rag. A trace of petroleum jelly will protect the steel from rust.

(*g*) *Painted woodwork* (thermometer screens etc.).   Woodwork should be brushed clean whenever necessary, and at stations affected by soot or smoke a thorough cleaning with soap and water should be carried out once a month. Detergent and water may also be used.

(*h*) *Painted surfaces liable to inking.*   The ink should be removed while wet with a damp cloth. Older stains can be removed by the application of a small quantity of whiting applied with a damp cloth. Methylated spirit may be used with the whiting if there is no risk of this getting on to lacquered brass or polished woodwork. Inhibisol or similar product may also be used.

(*i*) *Chromium-plated parts.*   Use a soft dry cloth. Do not apply metal polish.

1.10.3. Housing of loose meteorological instruments and equipment.   Stations making a full set of observations may have the following additional items which are not permanently installed but must be readily available for use in the enclosure:

(*a*) rain measure,

(*b*) grass minimum thermometer,

(*c*) bare-soil minimum thermometer,

(*d*) concrete-slab minimum thermometer, and

(*e*) extra rain-gauge bottles.

If the enclosure is considered safe from petty theft or interference, the rain measure and the three minimum thermometers may be left out of doors. Otherwise loose equipment must be brought back into the office after use.

If left out of doors, the rain measure must be inverted over a short wooden post or dowel which should be driven into the ground conveniently close to the rain-gauge. The diameter of the post or dowel should be only very slightly less than the diameter of the rain measure. If kept within the office, the rain measure should be retained inverted on a suitable holder.

Two of the additional minimum thermometers may be stored in the screen after reading. If the grass minimum thermometer has been exposed on attached rubber supports, the thermometer (with the supports attached) can simply be placed in the front well of the floor of the screen. The bare-soil minimum or concrete minimum thermometer may then be stored by resting it in the U-shaped slots of those rubber supports; this will ensure that both thermometers are stored at the required angle of 2° to the horizontal, with the bulb end lower than the stem. Where the special rubber supports are not used, the thermometers can be stored in the screen in a near-vertical or sloping position, with the bulb end in the well provided. If, for any reason, the thermometers cannot be stored in the screen, they should be brought indoors and stored in their appropriate containers in a near-vertical position, bulb end down.

With the exception of the minimum thermometers, as detailed above, the thermometer screen must not be used for the stowage of any extraneous equipment.

1.11. TABULATION AND CUSTODY OF AUTOGRAPHIC RECORDS

Considerable official use is made of the tabulations from autographic records in answering enquiries and for investigations. Moreover, the original anemograms, barograms, thermograms, hygrograms or record of rainfall charts are often required for special study, either in the Meteorological Office or by individuals in other organizations, or for reproduction and publication. It is of the utmost importance that every care should be taken to ensure that the instruments are working properly and that the records are satisfactory. Charts should be kept clean, with distinct unbroken traces and accurate time marks so that reliable tabulations may be prepared (see also 1.1.4). Each meteorological office in the United Kingdom equipped with the necessary instruments makes tabulations, on the appropriate forms, of anemograms, record of rainfall charts and sunshine records obtained at the station.

Photocopies of tabulations are often supplied in answer to enquiries. It is therefore essential that tabulations should be completed in black or blue-black permanent ink and the use of blue or other light shades of ink should be avoided. Inks which fade with time must not be used.

Serious faults on the more complicated autographic instruments (e.g. anemographs) which cannot be dealt with locally should be referred by Meteorological Office stations to their Area or Regional Maintenance Centre for advice, and by other stations to their parent station or administering authority as appropriate.

CHAPTER 2

CLOUDS

A cloud is a hydrometeor (see 4.2.1) in suspension in the free air and usually not touching the ground. A cloud is composed of minute particles of water or ice, or both. It may also contain larger particles of water or ice and non-aqueous liquid particles or solid particles such as those present in fumes, smoke or dust.

2.1. HISTORICAL NOTE

The classification of clouds is based upon that originated by Luke Howard in 1803, namely cirrus, the thread cloud; cumulus, the heap cloud; stratus, the flat cloud or level sheet; and nimbus, the rain cloud. The details of a more precise classification occupied the attention of meteorologists in many countries during the latter part of the nineteenth century. A classification was agreed at the International Meteorological Conference in Munich in 1891, and in 1896 the first edition of the *International cloud atlas* appeared.

An International Commission for the Study of Clouds was set up in 1921 and produced the *International atlas of clouds and of types of skies* in 1932, preceded by an abridged edition in 1930 for the use of observers to meet the requirements of coding changes. New editions of both publications followed in 1939.

In the 1932 atlas, concepts were introduced that were of very great importance. Clouds in the first atlas had been regarded rather as fixed entities with no individual life history but in the 1932 atlas were acknowledged as dynamic features of the sky, to be considered both in relation to all other clouds present and to their stage of development in time. This radical change in the way clouds should be viewed led to changes in nomenclature which were reflected in the international codes for reporting cloud types. In the atlas an attempt was made to give examples not only of the basic cloud classification but also of the sky types (to which numbers are given in the reporting codes).

A new *International cloud atlas* was published in two volumes in 1956 by the World Meteorological Organization (WMO). The contents of Volume I differed materially from the 1939 atlas. The grouping of clouds in 'cloud families' was abandoned, the classification into genera was maintained, but some details in the definitions were modified. The species and varieties were extended and considerably modified, and a new concept, that of 'mother-cloud' (see 2.3.5), was introduced.

A revised edition of Volume I of the WMO atlas was published in 1975. This chapter is based very closely on that edition from which the listed definitions have been taken.

18

2.2. OBSERVATION OF CLOUDS

For a synoptic report the observer should identify all the cloud forms present, estimate cloud amounts, and estimate or measure the height of the cloud bases. He should also study the sky as a whole to assess its general appearance. (For a climatological report the observer usually observes only the total cloud amount.)

The observer needs to keep an almost continuous watch on the sky to be able to correctly identify clouds which continuously evolve and change. It is not possible, for example, to identify mother-clouds if no watch has been maintained during their evolution.

The observer should avoid looking directly at the sun since this could cause serious impairment of sight. A cloud-observing visor should be used to detect very thin cirrus clouds which are barely visible against the blue of the sky, and to observe all clouds when there is haze. It must be stressed that the visor does not protect the eyes when looking directly at the sun.

At night the observer should allow two or three minutes for his eyes to become adapted to darkness before making the observation which should be made from a dark place as far as possible from lights, especially when the atmosphere is hazy.

Pictorial guides and coding instructions are given in *Cloud types for observers*. Full details of relevant codes and specifications are given in the *Handbook of weather messages*, Parts II and III.

2.2.1. Observations of clouds from mountain stations. The procedure for observing clouds from mountain stations is the same as that for low-level stations when the mountain station is at a level lower than that of the base of the cloud.

When clouds are observed below station level they should be indicated separately. A description should be given of the upper surface of such clouds; features such as a flat or an undulated surface, or the presence of towering cumuliform clouds above the top of the layer, should be recorded. In estimating the cloud amount, the places where mountains penetrate a patch, sheet or layer of clouds are considered as covered with clouds.

2.2.2. Appearance of clouds. The appearance of a cloud is best described mainly in terms of the dimensions, shape, structure and texture. However, two other factors affect the appearance of a cloud, namely 'luminance' and 'colour', which are briefly noted below.

2.2.2.1. The luminance of a cloud is determined by the light reflected, scattered and transmitted by its constituent particles. This light mainly comes direct from the sun or moon or from the sky; it may also come from the surface of the earth, and is particularly strong when sunlight or moonlight is reflected by ice-fields or snow-fields. Haze affects the luminance of cloud. On a moonlit night, clouds are visible when the moon is more than a quarter full.

Ice-crystal clouds are usually more transparent than water-droplet clouds because of their limited vertical thickness and the sparseness of the ice particles but sometimes, when illuminated from behind, show marked contrasts in luminance; they are, however, brilliantly white in reflected light.

2.2.2.2. The colour of clouds depends primarily on the colour of the incident light. Haze between the observer and cloud may modify colour; it tends, for instance, to make distant clouds look yellow, orange or red. Cloud colours are also influenced by special luminous phenomena which are described in Chapter 11.

When the sun is sufficiently high above the horizon, clouds (or portions of cloud) which chiefly diffuse light from the sun are white or grey. Parts which receive light mainly from the blue sky are bluish-grey. When the illumination by the sun and sky is extremely weak, the clouds tend to take the colour of the surface below them. Cloud colours vary with the height of the cloud and its relative position with regard to observer and sun.

At night, clouds usually appear black to grey unless illuminated by the moon when they present a whitish appearance. Special illumination (fires, street lighting, polar aurorae, etc.) may, however, give a marked colouring to certain clouds.

2.2.3. Height, altitude and vertical extent of cloud.

(a) Height of a cloud feature, e.g. the base or top of a cloud, is the vertical distance from the place of observation (which can be on a hill or mountain) to the level of the cloud feature. The height of the cloud base is an important factor in determining the cloud type.

(b) Altitude of a cloud feature, e.g. the base or top of a cloud, is the vertical distance measured from mean sea level to the level of the cloud feature.

(c) Vertical extent of a cloud is the vertical distance between the level of its base and that of its top.

Surface observers generally report height and aircraft observers report altitude. In order to avoid possible confusion in giving reports of cloud height in plain language, it is Meteorological Office practice to add the phrase 'above ground level' which is routine for actual cloud-height reports (but see 2.2.1 for mountain stations).

2.2.3.1. *Cloud levels*,   Observations have shown that cloud levels (with the exception of nacreous and noctilucent clouds, described in Chapter 11) vary over a range of altitudes from near sea level to perhaps 18 kilometres (60 000 feet) in the tropics, 14 km (45 000 ft) in middle latitudes, and 8 km (25 000 ft) in polar regions. By convention, the vertical extent of the atmosphere in which clouds are usually present is divided into three layers: high, medium and low. Each layer is defined by a range of altitudes between which clouds of certain types occur most frequently. The layers overlap and the approximate limits vary with the latitude as shown in the table below. Some clouds usually have their bases in the low level but may reach into the medium or high levels.

Level	Polar regions	Temperate regions	Tropical regions
High	3–8 km (10 000–25 000 ft)	5–14 km (16 500–45 000 ft)	6–18 km (20 000–60 000 ft)
Medium	2–4 km (6500–13 000 ft)	2–7 km (6500–23 000 ft)	2–8 km (6500–25 000 ft)
Low	From the earth's surface to 2 km (6500 ft)	From the earth's surface to 2 km (6500 ft)	From the earth's surface to 2 km (6500 ft)

When the height of the base of a particular cloud is known, the use of levels may assist the observer to identify the cloud.

2.3. CLASSIFICATION OF CLOUDS

This section details a classification of the characteristic forms of clouds. The definitions given apply, unless otherwise specified, to observations made under the following ideal conditions:

(*a*) The observer is at the earth's surface, either on land in areas without mountainous relief, or at sea.

(*b*) The air is clear; no obscuring phenomena such as fog, haze, dust, smoke, etc. are present.

(*c*) The sun is sufficiently high to give normal daylight conditions.

(*d*) The clouds are so high above the horizon that effects of perspective are negligible.

It will be necessary to adapt the definitions to other conditions. In many cases this can easily be done; for example, by night when the moon is in its brighter phases it may play, with regard to the illumination of clouds, a role analogous to that of the sun.

Following the *International cloud atlas*, the cloud forms are classified in terms of 'genera', 'species' and 'varieties'. A description of these terms follows, with an explanation of the subdivisions which can be used to define a particular cloud formation in more detail.

2.3.1. Genera. Cloud forms are divided into 10 main groups, called genera. A given cloud belongs to only one genus. The genera, together with their accepted form of abbreviation, are:

High	Medium	Low
Cirrus (Ci)	Altocumulus (Ac)	Stratus (St)
Cirrocumulus (Cc)	Altostratus (As)	Stratocumulus (Sc)
Cirrostratus (Cs)	Nimbostratus (Ns)	Cumulus (Cu)
	(Ns usually extends	Cumulonimbus (Cb)
	to other levels)	

The following definitions of genera are limited to a description of these 10 main types and the essential characteristics necessary to distinguish a given genus from genera having a similar appearance. It should be noted that, so far as reporting procedures are concerned, it is the main genera that are important, whilst a knowledge of the subdivisions (described in subsequent paragraphs) can assist in coding and in understanding the evolution and transformation of the types of cloud which are observed.

(1) *Cirrus.* Detached clouds in the form of white delicate filaments, or white or mostly white patches or narrow bands. These clouds have a fibrous (hair-like) appearance, or a silky sheen, or both.

Cirrus is whiter than any other cloud in the same part of the sky. With the sun on the horizon, it remains white whilst other clouds are tinted yellow or orange, but as the sun sinks below the horizon the cirrus takes on these colours too and the lower clouds become dark or grey. The reverse is true at dawn when the cirrus is the first to show coloration. It is of a mainly fibrous or silky appearance with no cloud elements, grains or ripples which distinguishes it from cirrocumulus. Cirrus is of a discontinuous structure and of limited horizontal extent which is the distinction between it and cirrostratus. Dense cirrus patches or tufts of cirrus may contain ice crystals large enough to fall to give trails of some vertical extent. If these crystals melt they appear greyish. (See Plates I and II.)

(2) *Cirrocumulus*. Thin white patch, sheet or layer of cloud without shading, composed of very small elements in the form of grains, ripples, etc. merged or separate, and more or less regularly arranged; most of the elements have an apparent width of less than 1°.

This cloud often forms as a result of the transformation of cirrus or cirrostratus. It is rippled and subdivided into very small cloudlets without any shading. It is composed almost exclusively of ice crystals and can include parts which are fibrous or silky in appearance but these do not collectively constitute its greater part. Cirrocumulus is usually associated with cirrus or cirrostratus, particularly in high and middle latitudes. Care should be taken when a layer of altocumulus is dispersing that the small elements on the edge of the sheet are not confused with cirrocumulus. (See Plates II and III.)

(3) *Cirrostratus*. Transparent whitish cloud veil of fibrous or smooth appearance, totally or partly covering the sky, and generally producing halo phenomena.

This cloud usually forms a veil of great horizontal extent, without structure and of a diffuse general appearance. It is composed almost entirely of ice crystals and can be so thin that the presence of a halo may be the only indication of its existence. Its apparent motion is slow, as are changes in its appearance and thickness. Cirrostratus, when not covering the whole sky, may have a clear-cut edge but more often it is fringed with cirrus. A veil of haze may sometimes have the appearance of cirrostratus but the haze veil is opalescent or has a dirty yellowish or brownish colour. Shadows will normally continue to be cast when the sun is shining through cirrostratus, at least when the sun is high; when the sun is less than about 30° above the horizon the relatively longer path through a cirrostratus veil may reduce the light intensity so much that shadows do not form. (See Plates III and IV.)

(4) *Altocumulus*. White or grey, or both white and grey, patch, sheet or layer of cloud, generally with shading, and composed of laminae, rounded masses, rolls, etc. which are sometimes partly fibrous or diffuse and which may or may not be merged; most of the regularly arranged small elements usually have an apparent width of between 1 and 5°.

This genus can be confused with cirrocumulus and stratocumulus. If the cloud elements exhibit any shading, the cloud is altocumulus not cirrocumulus even if the elements have an apparent width of less than 1°. Clouds without shading are altocumulus if most of the regularly arranged elements, when observed at more than 30° above the horizon, have an apparent width of 1 to 5°. (Under similar conditions of observation, stratocumulus elements will have an apparent width of over 5°.) Altocumulus sometimes produces descending trails of fibrous appearance (virga), and coronae and irisation are often observed in thin parts of the cloud. (See Plates IV, V and VI.)

(5) *Altostratus*.   Greyish or bluish cloud sheet or layer of striated, fibrous or uniform appearance, totally or partly covering the sky, and having parts thin enough to reveal the sun at least vaguely, as if through ground glass. Altostratus does not show halo phenomena.

Altostratus prevents objects on the ground from casting shadows. If the presence of the sun or moon can be detected, this indicates altostratus rather than nimbostratus. The former has a less uniform base. If it is very thick and dark, differences in thickness may cause relatively light patches between darker parts, but the surface never shows real relief, and the striated or fibrous structure is always seen in the body of the cloud. At night, if there is any doubt as to whether it is altostratus or nimbostratus when no rain or snow is falling, then, by convention, it is called altostratus. Altostratus is never white, as thin stratus may be when viewed more or less towards the sun. (See Plate VI and VII.)

(6) *Nimbostratus*.   Grey cloud layer, often dark, the appearance of which is rendered diffuse by more or less continuously falling rain or snow which, in most cases, reaches the ground. It is thick enough throughout to blot out the sun. Low ragged clouds frequently occur below the layer, with which they may or may not merge.

Nimbostratus generally forms from thickening altostratus or when a well-developed cumulonimbus thickens and spreads out. Note that if hail, thunder or lightning are produced by the cloud it is called cumulonimbus. Nimbo stratus is a thick dense cloud and can be distinguished from thick stratus by the type of precipitation it produces (see 2.3.6). It is also generally an extensive cloud, the base of which can be partially or totally hidden by ragged clouds (pannus), and care must be taken not to confuse these with the base of the nimbostratus. (See Plate VIII.)

(7) *Stratus*.   Generally grey cloud layer with a fairly uniform base, which may give drizzle, snow or snow grains. When the sun is visible through the cloud its outline is clearly discernible. Stratus does not produce halo phenomena except possibly at very low temperatures. Sometimes stratus appears in the form of ragged patches.

Stratus may develop from stratocumulus when the cloud base of the latter becomes lower or loses its relief or apparent subdivisions. The lifting of a fog layer due to warming or an increase in the wind speed is another common mode of formation. During formation or dissipation it can appear as more or less joined fragments with varying luminance as stratus fractus but this stage is usually fairly short. It is usually composed of small water droplets and, if the cloud is thin, these can give rise to a corona. (See Plates VIII and IX.)

(8) *Stratocumulus*.   Grey or whitish, or both grey and whitish, patch, sheet or layer of cloud which almost always has dark parts, composed of tesselations, rounded masses, rolls, etc. which are non-fibrous (except for virga) and which may or may not be merged; most of the regularly arranged small elements have an apparent width of more than 5°. (If a centimetre rule is held at arm's length, each centimetre will approximate to 1°.)

Stratocumulus may sometimes be confused with altocumulus. If most of the regularly arranged elements, when observed at an angle of more than 30° above the horizon, have an apparent width of more than 5°, the cloud is stratocumulus. Generally stratocumulus is composed of water droplets and, when it is not very thick, a corona or irisation is sometimes seen. However, in very cold weather, it may produce abundant ice-crystal virga. (See Plates IX and X.)

(9) *Cumulus*. Detached clouds, generally dense and with sharp outlines, developing vertically in the form of rising mounds, domes or towers, of which the bulging upper part often resembles a cauliflower. The sunlit parts of these clouds are mostly brilliant white; their base is relatively dark and nearly horizontal. Sometimes cumulus is ragged.

Cumulus clouds are detached although, when viewed from a distance, they may appear to have merged owing to the effect of perspective. Owing to their generally great vertical extent, cumulus may spread out and form stratocumulus or altocumulus cumulogenitus or, alternatively, penetrate existing layers of altocumulus or stratocumulus. Providing the cumuliform clouds remain detached from one another, they are still called cumulus. It can sometimes happen that a very large precipitating cumulus directly above an observer will exhibit none of the features normally associated with this genus, and it may be confused with altostratus or nimbostratus. Cloud evolution and the nature of the precipitation (see 2.3.6) can be of assistance in this case. (See Plates X, XI and XII.)

(10) *Cumulonimbus*. Heavy and dense cloud, of considerable vertical extent, in the form of a mountain or huge towers. At least part of its upper portion is usually smooth, or fibrous or striated, and nearly always flattened; this part often spreads out in the shape of an anvil or vast plume.

Under the base of this cloud, which is often very dark, there are frequently low ragged clouds either merged with it or not, and precipitation sometimes in the form of virga.

Cumulonimbus clouds normally develop from large cumulus but they can also do so from stratocumulus castellanus or altocumulus castellanus. When they cover a large expanse of sky the under surface can present the appearance of nimbostratus. The character of the precipitation may be of assistance in identifying the cloud. Cumulonimbus gives showers, very often quite heavy for comparatively short periods of time. If hail, thunder or lightning are observed then, by convention, the cloud is cumulonimbus. The evolution of the cloud can also aid identification. The change from large cumulus with domed tops and a hard outline (produced by water drops) to a top with a softer fibrous outline (produced by ice crystals) marks the change from cumulus to cumulonimbus. This genus may be described as a 'cloud factory'; it may produce extensive thick patches of cirrus spissatus, altocumulus, altostratus or stratocumulus by the spreading out of its upper portions. The spreading of the highest part usually leads to the formation of an anvil; if the wind increases with height, the upper portion of the cloud is carried downwind in the shape of a half anvil or vast plume. (See Plate XII.)

2.3.2. Species. Peculiarities in the shape of clouds and differences in their internal structure have led to the subdivision of most of the cloud genera into species. Altostratus and nimbostratus are the only genera which are not so divided. A cloud identified as of any particular genus may bear the name of one species only. The species are mutually exclusive but when several clouds of one particular genus are present simultaneously these clouds need not all belong to the same species. Certain species are common to several genera; for example, a profile resembling the shape of an almond or a lens is frequently seen in clouds of the genera cirrocumulus and altocumulus (and rarely in stratocumulus); consequently 'lenticularis' is recognized as a species in each of these three genera. The full list of cloud species is given below.

A cloud need not be identified as belonging to any species. When for a cloud of a given genus none of the definitions relevant to the genus is applicable, no species is indicated.

Fractus (fra)
Clouds in the form of irregular and ragged shreds. This terms applies only to stratus and cumulus. (See Plate VIII.)

Nebulosus (neb)
A cloud like a nebulous veil or layer, showing no distinct details. This term applies mainly to stratus and cirrostratus. (See Plate IV and IX)

Stratiformis (str)
Clouds spread out in an extensive horizontal sheet or layer. This term applies to stratocumulus, altocumulus and, occasionally, to cirrocumulus. (See Plates IV, IX and X.)

Lenticularis (len)
Clouds having the shape of a lens or almond, often very elongated and usually with well-defined outlines; they occasionally show irisation (iridescence). Such clouds appear most often in cloud formations of orographic origin, but may also occur in regions without marked orography. This term applies mainly to stratocumulus, altocumulus and cirrocumulus. (See Plate V.)

Castellanus (cas)
Clouds which present, in at least some portion of their upper part, cumuliform protuberances in the form of turrets which generally give the clouds a crenellated appearance. The turrets, some of which are taller than they are wide, may be connected by a common base and seem to be arranged in lines. The castellanus character is especially evident when the clouds are seen from a distance. This term applies to stratocumulus, altocumulus, cirrus and cirrocumulus. (See Plate VI.)

Humilis (hum)
Cumulus clouds of only slight vertical extent; they generally appear flattened. (See Plate X.)

Mediocris (med)
Cumulus clouds of moderate vertical extent, the tops of which show fairly small protuberances. (See Plate XI.)

Congestus (con)
Cumulus clouds which are markedly sprouting and are often of great vertical extent; their bulging upper part frequently resembles a cauliflower. (See Plates XI and XII.)

Calvus (cal)

Cumulonimbus in which at least some of the protuberances of the upper part are beginning to lose their cumuliform outlines but in which no cirriform parts can be distinguished. Protuberances and sproutings tend to form a whitish mass with more or less vertical striations. (See Plate XII.)

Capillatus (cap)

Cumulonimbus characterized by the presence, mostly in its upper portion, of distinct cirriform parts of clearly fibrous or striated structure, frequently having the form of an anvil, a plume or a vast, more or less disorderly, mass of hair. (See Plate XII.)

Floccus (flo)

A species in which each cloud unit is a small tuft with a cumuliform appearance, the lower part of which is more or less ragged and often accompanied by virga. This term applies to altocumulus, cirrus and cirrocumulus. (See Plate V.)

Fibratus (fib)

Detached clouds or a thin cloud veil, consisting of nearly straight or more or less irregularly curved filaments which do not terminate in hooks or tufts. This term applies mainly to cirrus and cirrostratus. (See Plates I and III.)

Spissatus (spi)

Cirrus of sufficient optical thickness to appear greyish when viewed towards the sun.

Uncinus (unc)

Cirrus often shaped like a comma, terminating at the top in a hook, or in a tuft whose upper part is not in the form of a rounded protuberance. (See Plate I.)

2.3.3. Varieties. Special features in appearance and degree of transparency of clouds have led to the concept of varieties. For example, a cloud with a definite wave characteristic would be termed 'undulatus'. The varieties, apart from 'translucidus' and 'opacus', are not mutually exclusive, so a particular cloud may bear the names of different varieties. Certain varieties may occur in clouds of several genera. The full list of cloud varieties is given below, but a cloud need not be identified as being of any particular variety.

The varieties duplicatus, intortus, lacunosus, radiatus, undulatus and vertebratus refer to the appearance; the varieties opacus, perlucidus and translucidus refer to the degree of transparency.

Duplicatus (du)

Superposed cloud patches, sheets or layers at slightly different levels, sometimes partly merged. This term applies mainly to stratocumulus, altocumulus, altostratus, cirrus and cirrocumulus.

Intortus (in)

Cirrus, the filaments of which are very irregularly curved and often seemingly entangled in a capricious manner. (See Plate II.)

Plate I

Cirrus (species fibratus). Fine white filaments nearly straight without hooks or tufts.

Cirrus (species uncinus). White filaments terminating in a hook or a tuft and shaped like a comma in places.

Plate II

Cirrus (variety intortus). The filaments are entangled in a capricious manner.

Cirrocumulus (variety lacunosus). The cloud is without shading and is composed of ripples, grains, etc. which identifies it as cirrocumulus. The variety lacunosus is seen at the top of the photograph.

Plate III

Cirrocumulus (variety undulatus). The cloud elements in this broken sheet are both separate and merged in places

Cirrostratus. The veil of fibrous cloud typical of this genera is illustrated here, with cirrus fibratus at the top of the photograph.

Plate IV

Cirrostratus (species nebulosus). The appearance of the 22-degree radius halo indicates that the cloud is composed of ice crystals.

Altocumulus (species stratiformis, varieties translucidus and perlucidus). The elements in altocumulus are larger than those in cirrocumulus and exhibit shading. The cloud depicted is sufficiently translucent to reveal the position of the sun, hence translucidus. The spaces between the cloud elements signify the variety perlucidus.

Plate V

Altocumulus (species lenticularis). The patches of cloud are each in the shape of an elongated lens or almond and exhibit fairly well-defined edges.

Altocumulus (species floccus). These clouds resemble very small, more or less ragged cumulus; they are often accompanied by fibrous trails (virga).

Plate VI

Altocumulus (species castellanus). Altocumulus composed of turrets which appear to be arranged in lines. The turrets generally have a common horizontal base.

Altocumulus and altostratus (variety opacus with supplementary feature mamma). The sky as a whole featured both genera and this photograph illustrates the well-formed protuberances known as mamma.

Plate VII

Altostratus (variety translucidus). With thin altostratus the sun (or moon) loses its sharp outline, as though seen through ground glass. Note the absence of halo, indicating that the cloud is composed of water drops.

Altostratus (variety opacus). Although there is a broken layer of stratocumulus present in this photograph, the appearance of a thick layer of altostratus is well illustrated.

Plate VIII

Nimbostratus. The generally grey appearance of this type of cloud, which lacks any well-defined base, is illustrated here. Nimbostratus is thick enough to blot out the sun. Continuous rain or snow, in most cases reaching the ground, accompanies the cloud. Pannus is frequently present.

Stratus (species fractus). Broken stratus cloud in the form of irregular shreds with a ragged appearance.

Plate IX

Stratus (species nebulosus, variety opacus). The diffuse irregular base of this cloud is shown up against the mountain.

Stratocumulus (species stratiformis, variety perlucidus). The elements and shading of this genera are illustrated.

Plate X

Stratocumulus (species stratiformis, variety undulatus). The rolls in this sheet are well separated.

Cumulus (species humilis). A typical example of an early summer-day development of small cumulus. As daytime temperatures rise, their vertical development may greatly increase.

Plate XI

Cumulus (species congestus). The cloud has marked vertical development but the outline is still clear-cut which serves to distinguish it from cumulonimbus species calvus.

Cumulus (species mediocris). A line or street of cumulus showing progressive vertical development. The base is still well-defined with no shower activity obvious.

Plate XII

Cumulus (species congestus) with cumulonimbus calvus. A cumulus cloud of the species congestus, with a bulging upper part resembling a cauliflower, dominates the photograph. To the left can be seen the top of a cumulonimbus calvus cloud, the summit of which has lost its sharp outline.

Cumulonimbus (species capillatus). The distinct cirriform head of a cumulonimbus cloud; the description of capillatus as a disordered mass of hair is particularly well illustrated.

Plate XIII

METEOROLOGICAL OFFICE CLOUD SEARCHLIGHT

METEOROLOGICAL OFFICE ALIDADE

Plate XIV

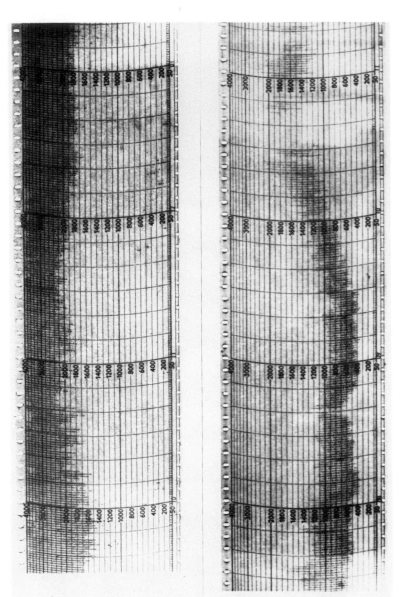

CLOUD-BASE RECORDER CHARTS

Upper chart: continuous layer of cloud with base varying between 1400 and 2200 ft. Lower chart: cloud base about 500 ft, lifting with time to about 1300 ft and breaking.

Plate XV

METEOROLOGICAL OFFICE VISIBILITY METER Mk 2

Plate XVI

SOFT RIME, JANUARY 1979
See page 62

GLAZE, FEBRUARY 1979
See page 63

Lacunosus (la)

Cloud patches, sheets or layers, usually rather thin, marked by more or less regularly distributed round holes, many of them with fringed edges. Cloud elements and clear spaces are often arranged in a manner suggesting a net or a honeycomb. This term applies mainly to altocumulus and cirrocumulus (see Plate II); it may also apply, though rarely, to stratocumulus.

Radiatus (ra)

Clouds showing broad parallel bands or arranged in parallel bands which, owing to the effect of perspective, seem to converge towards a point on the horizon or, when the bands cross the whole sky, towards two opposite points on the horizon, called 'radiation point(s)'. This term applies mainly to strato-cumulus, cumulus, altocumulus, altostratus and cirrus.

Undulatus (un)

Clouds in patches, sheets or layers showing undulations. These undulations may be observed in fairly uniform cloud layers or in clouds composed of elements, separate or merged. Sometimes a double system of undulations is in evidence. This term applies mainly to stratus, stratocumulus, altostratus, altocumulus, cirrostratus and cirrocumulus. (See Plates III and X.)

Vertebratus (ve)

Clouds whose filaments are arranged in a manner suggestive of vertebrae, ribs or a fish skeleton. This term applies mainly to cirrus.

Opacus (op)

An extensive cloud patch, sheet or layer, the greater part of which is sufficiently opaque to mask the sun or moon completely. This term applies to stratus, stratocumulus, altostratus and altocumulus. (See Plates VI, VII and IX.)

Perlucidus (pe)

An extensive cloud patch, sheet or layer, with distinct but sometimes very small spaces between the elements. The spaces allow the sun, the moon, the blue of the sky or overlying clouds to be seen. The variety perlucidus may be observed in combination with the varieties translucidus or opacus. This term applies to stratocumulus and altocumulus. (See Plates IV and IX.)

Translucidus (tr)

Clouds in an extensive patch, sheet or layer, the greater part of which is sufficiently translucent to reveal the position of the sun or moon. This term applies to stratus, stratocumulus, altostratus and altocumulus. (See Plates IV and VII.)

2.3.4. Supplementary features and accessory clouds. The indication of genus, species and variety is not always sufficient to describe the cloud com-pletely. The cloud may sometimes have supplementary features or may be accompanied by other, usually smaller, clouds which are known as accessory clouds and which may be separate or partly merged with the main cloud. Supplementary features and accessory clouds may occur at any level of the cloud, above it or below it. One or more supplementary features or accessory clouds may occur simultaneously with the same cloud.

Definitions of the supplementary features and accessory clouds are given below.

(a) Supplementary features:

Virga (vir)

Vertical or inclined trails of precipitation (fallstreaks), attached to the under surface of a cloud, which do not reach the earth's surface. This supplementary feature occurs mostly with stratocumulus, cumulus, cumulonimbus, altostratus, altocumulus, nimbostratus and cirrocumulus.

Praecipitatio (pra)

Precipitation (rain, drizzle, snow, hail, etc.) falling from a cloud and reaching the earth's surface. (Although precipitation is normally considered a hydrometeor (see 4.2.1), it is treated here as a supplementary feature because it appears as an extension of the cloud.) This supplementary feature is mostly encountered with stratus, stratocumulus, cumulus, cumulonimbus, altostratus and nimbostratus. (See Plate XII.)

Mamma (mam)

Hanging protuberances, like udders, on the under surface of a cloud. This supplementary feature occurs mostly with stratocumulus, cumulonimbus, altostratus, altocumulus, cirrus and cirrocumulus. (See Plate VI.)

Arcus (arc)

A dense horizontal roll, with more or less tattered edges, situated on the lower front part of certain clouds and having, when extensive, the appearance of a dark menacing arch. This supplementary feature occurs with cumulonimbus and, less often, with cumulus.

Incus (inc)

The upper portion of a cumulonimbus spread out in the shape of an anvil, with a smooth fibrous or striated appearance.

Tuba (tub)

Cloud column or inverted cloud cone (funnel cloud) protruding from a cloud base; it constitutes the cloudy manifestation of a more or less intense vortex. This supplementary feature occurs with cumulonimbus and, less often, with cumulus. The diameter of the cloud column, which is normally of the order of 10 metres, may in certain regions occasionally reach some hundreds of metres (see definition of spouts in 4.2.1.19 on page 63).

(b) Accessory clouds:

Pannus (pan)

Ragged shreds, sometimes constituting a continuous layer, situated below another cloud and sometimes attached to it. This accessory cloud occurs mostly with cumulus, cumulonimbus, altostratus and nimbostratus.

Pileus (pil)

An accessory cloud of small horizontal extent in the form of a cap or hood above the top, or attached to the upper part, of a cumuliform cloud which often penetrates it. Pileus clouds may sometimes be observed one above the other. Pileus occurs principally with cumulus and cumulonimbus.

Velum (vel)

An accessory cloud veil of great horizontal extent, close above or attached to the upper part of one or several cumuliform clouds which often pierce it. Velum occurs principally with cumulus and cumulonimbus.

2.3.5. Mother-clouds.   Clouds may form from other clouds, called mother-clouds, in two ways.

Firstly, a part of a cloud may develop and more or less pronounced extensions may form. These extensions, whether attached to the mother-cloud or not, may become clouds of a genus different from that of the mother-cloud. They are then classified as the appropriate genus, followed by the genus of the mother-cloud with the addition of the suffix 'genitus' (e.g. stratocumulus cumulogenitus).

Secondly, the whole or a large part of a cloud may undergo complete internal transformation, thus changing from one genus into another. The new cloud is then classified as the appropriate genus, followed by the genus of the mother-cloud with the addition of the suffix 'mutatus' (e.g. stratus strato-cumulomutatus). The internal transformation of clouds should not be confused with changes in the appearance of the sky resulting from the relative movement of clouds and observer.

No mention should be made of mother-clouds if there is any doubt concerning either the origin of the observed clouds ('genitus') or the manner of their development ('mutatus').

2.3.6. Aid to cloud identification.   If, especially during the hours of darkness, an observer has difficulty in identifying the cloud types following the onset of precipitation, the table below can be used as a guide to the cloud types which may be present from the nature of the precipitation.

Precipitation	Cloud type					
	As	Ns	Sc	St	Cu*	Cb*
Rain	+	+	+		+	+
Drizzle			+			
Snow	+	+	+	+	+	+
Snow pellets			+		+	+
Hail						+
Small hail						+
Ice pellets	+	+				
Snow grains				+		

*Showery precipitation
Very occasionally rain or snow may reach the earth's surface from altocumulus castellanus.

2.4. CLOUD AMOUNT

The total cloud amount, or total cloud cover, is the fraction of the celestial dome covered by all clouds visible. The assessment of the total amount of cloud therefore consists in estimating how much of the total apparent area of the sky is covered with cloud. The international unit for reporting cloud amount is the okta or eighth of the sky and estimates are made in this unit, with a special qualification for cloud amounts of less than 1 okta and greater than 7 oktas. See code figures 1 and 7 in 2.4.1 below.

The site used when estimating cloud amount should be one which commands the widest possible view of the sky and the observer should give equal weight to the areas overhead and those at the lower angular elevations. On occasions when the clouds are very irregularly distributed it is useful to consider the sky in separate quadrants divided by diameters at right angles to each other. The sum of the estimates for each quadrant is then taken as the total for the whole sky.

2.4.1. Complete range of cloud amounts.

Code figure	Cloud amount
0	None (sky completely cloudless)
1	1 okta or less, but not zero
2	2 oktas
3	3 oktas
4	4 oktas
5	5 oktas
6	6 oktas
7	7 oktas or more, but not 8
8	8 oktas (sky completely covered)
9	Sky obscured or cloud amount cannot be estimated.

Code figure 9 is reported when either the sky is invisible owing to fog, falling snow, etc. or the observer cannot estimate the amount owing to darkness or extraneous lighting. On moonless nights it should usually be possible to estimate the total amount by reference to the proportion of the sky in which the stars are dimmed or completely hidden by clouds, although haze alone may blot out stars near the horizon.

2.4.2. Partial cloud amounts.

The observer also has to estimate how many eighths of the sky would be covered by each individual type of cloud as if it alone was the only cloud type in the sky. There are times, for example, when a higher layer of cloud is partially obscured by lower clouds. In these cases an estimate can be made with comparative assurance of the extent of the upper cloud by watching the sky for a short time. Movement of the lower cloud relative to the higher should reveal whether the higher layer is completely covering the sky or has breaks in it. There are of course occasions of great difficulty, more especially at night, in making a reliable estimate; but previous observation of cloud development, general knowledge of cloud structure, and allowing sufficient time for the eyes to adjust to the dark will help the observer to achieve the best possible result. Access to reports from aircraft, if available, can also be of assistance.

It should be noted that the estimation of amount of each different type of cloud is made independently of the estimate of total cloud amount. The sum of separate estimates of partial cloud amounts often exceeds the total cloud amount, and often exceeds 8 eighths.

2.5. HEIGHT OF CLOUD BASE

In the United Kingdom the height of cloud base is recorded in feet but eventually it will be recorded in metres. However, to maintain the procedure which has been used throughout this book, feet 'equivalents' follow reference to heights in metres. The 'equivalents' are the values currently being recorded, but the metres figures do not represent the direct conversion of feet to metres but rather those values which would be recorded if cloud heights were recorded in metres.

Cloud height is difficult to estimate and also at times difficult to measure. At aerodromes the height of cloud base is often the most important fact reported by the meteorological observer. It is therefore essential that observers, especially at aerodromes, should use every means available to attain an accuracy of reporting within 10 per cent of the actual cloud heights up to 1500 m (5000 ft).

Where the station is not provided with equipment for measuring cloud height, the reported heights should be obtained by estimation, following the guidance given in 2.5.3.

Where measuring equipment is available, the height of the cloud base should be checked frequently to enable the height to be given to the accuracy shown above. At night the use of a cloud searchlight makes frequent checks practicable, and the height should be measured for each observation when cloud is present within the limits of the beam. Some stations are equipped with a Meteorological Office cloud-base recorder (see 2.5.6) which is used, day and night, for all observations. Separate instructions are available at such stations.

2.5.1. Definition of cloud base. The definition of the cloud base has given rise to much discussion. The World Meteorological Organization has defined it as 'that lowest zone in which the type of obscuration perceptibly changes from that corresponding to clear air or haze to that corresponding to water droplets or ice crystals'. Observers should regard the cloud base of a particular cloud type as (a) that level at which a pilot balloon is first seen to be obscured by the cloud (care being taken that lower clouds do not confuse the judgement), or (b) the lower limit of the patch of light formed on the cloud type by a searchlight beam, or the lowest value of the pen marks on a cloud-base recorder chart which are appropriate to the cloud type that is being assessed, or (c) when cloud height is estimated, the lowest level down to which cloud of the type reported is judged to extend. Difficulty will mainly arise when the cloud base is diffuse and irregular, as with stratus fractus of bad weather. If there are patches of such cloud coming down to say 30 metres (100 ft), although other portions of the same cloud are at a greater height, the height should be reported as 30 m (100 ft).

2.5.2. Datum for height of cloud base at aerodromes. The height of the cloud base is normally reported as its height above ground level where this datum (ground level) refers to the observation site or the site on which the rain-gauge is installed. On some aerodromes, however, the observation site is appreciably higher or lower than the published official aerodrome altitude. At

these aerodromes the cloud height is adjusted to relate to the official aerodrome altitude when the cloud base is 450 m (1500 ft) or below, and the
difference between the altitude of the aerodrome and the observing site is
12 m (40 ft) or more. The following table is used to determine the correction
to be applied to the observed cloud height, and the adjusted height is entered
in the Daily Register as the cloud base.

Difference between altitude of observation site and official aerodrome altitude		Correction to observed cloud height if observation site is			
		higher than official aerodrome altitude		lower than official aerodrome altitude	
m	*ft*	*m*	*ft*	*m*	*ft*
12–21	40–69	+15	+ 50	−15	− 50
22–30	70–99	+30	+100	−30	−100

2.5.3. Estimation of cloud height.   At stations not provided with measuring
equipment the values of cloud height can only be estimated. In mountain
areas the height of any cloud base which is lower than the tops of the hills or
mountains around the station can be estimated by comparison with the
heights of well-marked topographical features as given in the Ordnance Survey map of the district. It is useful to have, for permanent display, a diagram
detailing heights and bearings of hills and landmarks which might be useful in
estimating cloud height. Due to perspective the cloud may appear to be
resting on distant hills, and the observer must not necessarily assume this
reflects the height of the cloud over the observation site. In all circumstances
the observer must use his judgement, taking into consideration the form and
general appearance of the cloud.

The range of cloud-base heights above ground level applicable to various
genera of cloud over the British Isles is given in the table below and refers to
level country not more than 150 m (500 ft) above mean sea level. For observing sites at substantially greater heights, or stations on mountains, the height
of the base of low cloud above the station will often be less. For instance, at a
station in the British Isles between 300 and 450 m (1000 and 1500 ft), the
height indicated in the table below would be reduced by about 300 m
(1000 ft).

In countries with climates very different from that of the British Isles, and
especially in dry tropical conditions, cloud heights may depart substantially
from the ranges given. The differences may introduce problems of cloud
classification as well as increasing the difficulty of estimating the height. For
instance, reports of tropical cumulus clouds of an obviously convective origin,
with base well above 2400 m (8000 ft) or even as high as 3600 m (12 000 ft)
have been confirmed by aircraft observations. It is noteworthy that surface
observers frequently underestimate cloud heights to a very serious degree in
such cases. These low estimates may be due to two factors: either the observer expects cumulus cloud to be a 'low cloud' with its base below 2000 m
(6500 ft) and usually below 1500 m (5000 ft), or the atmospheric conditions
and the form of cloud may combine to produce an optical illusion.

When a direct estimate of cloud height is made at night, success depends
very greatly on the correct identification of the form of cloud. General
meteorological knowledge and a close watch on the weather are very important in judging whether a cloud base has remained substantially unchanged or

HEIGHTS OF THE BASE OF CLOUD GENERA
ABOVE GROUND LEVEL IN THE BRITISH ISLES

Cloud genera	Usual range of height of base*		Wider range of height of base sometimes observed, and other remarks	
	metres	*feet*	*metres*	*feet*
LOW				
Stratus	Surface–600	Surface–2000	Surface–1200	Surface–4000
Stratocumulus	300–1350	1000–4500	300–2000	1000–6500
Cumulus	300–1500	1000–5000	300–2000	1000–6500
Cumulonimbus	600–1500	2000–5000	300–2000	1000–6500
	kilometres			
MEDIUM			Nimbostratus is considered a medium cloud for synoptic purposes, although it can extend to other levels. Altostratus may thicken with progressive lowering of the base to become nimbostratus.	
Nimbostratus	Surface–3	Surface–10 000		
Altostratus ⎱ Altocumulus ⎰	2–6	6500–20 000		
HIGH			Cirrus from dissipating cumulonimbus may occur well below 6 km (20 000 ft) in winter. Cirrostratus may develop into altostratus.	
Cirrus ⎱ Cirrostratus ⎬ Cirrocumulus ⎰	6–12	20 000–40 000		

*For stations substantially over 150 metres (500 ft) above sea level, the base of low-level clouds will often be less.

has risen or fallen. A most difficult case, calling for great care and skill, occurs when a sheet of altostratus has covered the sky during the evening. Any gradual lowering of such a cloud sheet may be very difficult to detect but, as it descends, the base is rarely quite uniform and small contrasts can often be discerned on all but the darkest nights.

2.5.4. Measurement of cloud height by searchlight. In this method, illustrated in Figure 1, the angle of elevation E of a patch of light formed on the base of the cloud by a vertically directed searchlight beam is measured by alidade from a distant point. If L is the known horizontal distance in metres (feet) between the searchlight and the place of observation, then the height, h, in metres (feet) of the cloud base above the point of observation is given as $h = L \tan E$.

The optimum distance of separation between the searchlight and the place of observation is about 300 m (1000 ft). If the distance is much greater than this, the spot of light may be difficult to see; if it is much less, the accuracy of measuring height above about 600 m (2000 ft) suffers. A distance of 250–550 m (800–1800 ft) is usually acceptable.

The conversion of the angle of observation into a height in metres for a baseline of 300 m, and in feet for a baseline of 1000 ft, may be made by using Table II (a) or (b) on pages 206 and 207 respectively. Similar tables for any other baseline of L metres or L feet may be prepared by multiplying the values by $L/300$ for metres and by $L/1000$ for feet.

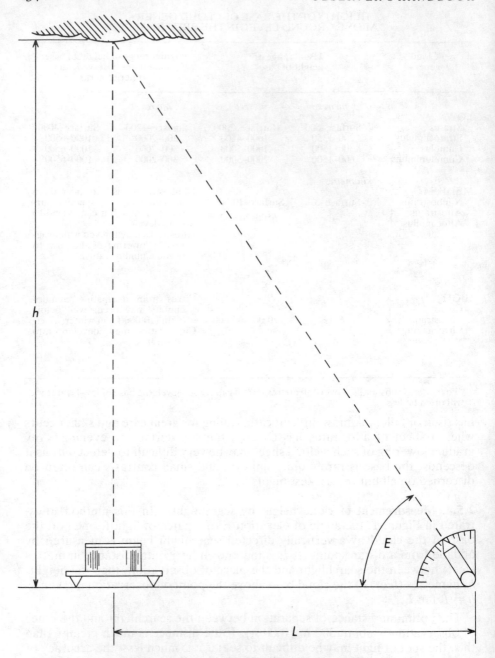

Figure 1.   Principle of cloud-height measurement with searchlight

The searchlight is to be used for all observations not made in daylight when the cloud is suitable. On a dark overcast night, cloud heights up to about 3000 m (10 000 ft) can be measured with acceptable accuracy (Figure 2(a)). The site chosen for the instrument must not be affected by extraneous lighting.

On many occasions some care will be necessary to ensure that cloud which terminates the beam at a particular instant represents the height of the lowest layer of cloud. In order to increase the chances of detecting small amounts of low cloud the beam should be observed over several minutes. If there are fragments of cloud below a main cloud layer and the beam can be seen illuminating the fragments (Figure 2 (b)) and the base of the main layer, then all heights should be determined.

When tenuous, ragged or fragmentary very low cloud is present the searchlight will produce a diffusely illuminated section in the upper part of its beam rather than a definite horizontal spot. Below this illuminated section the lower part of the beam is usually only faintly visible (Figure 2 (c)). During rain, however, the lower part of the beam will be more clearly visible because of the illumination of the raindrops. In poor visibility the lower part of the beam may be even brighter and the contrast between it and the illumination on the low cloud above not very well marked (Figure 2 (d)). In such cases the measurement of the height should be made by sighting the alidade (see 2.5.4.2) on the bottom of the brighter section of the beam, i.e. where the beam exhibits an appreciable increase in brightness with height (X in Figure 2 (c) and (d)).

In very heavy rain, sleet or snow it is sometimes impossible to get any bright spot on the searchlight beam (Figure 2 (e)), and the 'point' where the beam disappears in these conditions may provide no useful indication of the height of cloud.

Snow lying on the sloping glass window of the standard cloud searchlight may obliterate the beam. If the lamp is allowed to burn long enough, the heat generated will usually cause the snow to slide off.

At stations where frequent observations are made during the hours of darkness, the searchlight can be left switched on. Being left switched on has no marked effect on the life of the bulb and also prevents condensation taking place inside the searchlight. Where the searchlight is used only at, say, 3-hourly intervals it should be switched off as soon as an observation has been completed. If it is used only at fixed times the searchlight may be sited near a source of mains electricity and controlled by time-switches to avoid the high cost of cabling. This arrangement is unsuitable at aviation stations.

Spare searchlight bulbs must be kept readily available for immediate replacement of unserviceable bulbs.

2.5.4.1. *Standard cloud searchlight.*   The standard cloud searchlight used in the Meteorological Office (see Plate XIII) consists of a parabolic silvered-glass reflector, 40·6 cm (16 inches) in diameter, mounted in a strong cylindrical case, the base of which has three adjustable feet for levelling purposes. The top of the case is inclined to the horizontal and is covered by a piece of glass pressed tight between two packing rings. The metal ring which clamps the glass in position is provided with a slot at the lowest point to allow rain-water

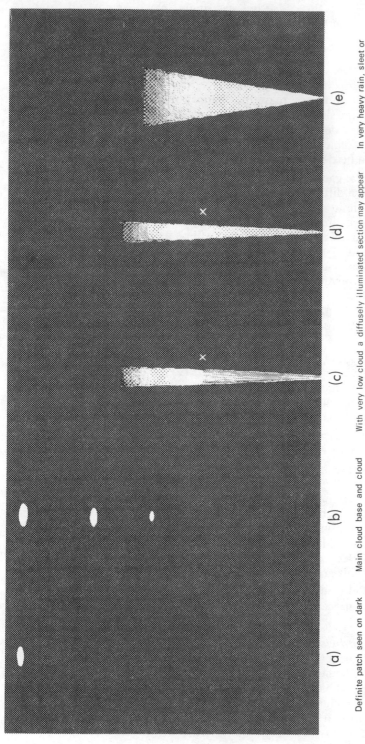

(a)

Definite patch seen on dark overcast nights. Cloud-base heights up to 3000 metres (10 000 ft) can be measured with the cloud searchlight.

(b)

Main cloud base and cloud fragments illuminated. The beam should be observed over several minutes and all heights determined.

(c) (d) (e)

With very low cloud a diffusely illuminated section may appear in the upper part of the beam with the lower part just visible (c). In rain the lower part of the beam is more clearly visible and in poor visibility may be even brighter (d). The alidade should be sighted on X, the bottom of the beam's brightest section.

In very heavy rain, sleet or snow no accurate reading may be possible if there is no definite bright spot on which to sight the alidade.

Figure 2.   Measurement of cloud base with searchlight

collected on the glass to run away. The lamp, which is usually 24 volts, 500 watts, works at a very high efficiency and consequently has a rather short life. There is adequate provision for focusing the lamp. Full instructions for setting up the instrument, adjusting the verticality of the beam, and general maintenance are available from the Operational Instrumentation Branch.

Whenever a bulb is changed the verticality of the beam must be checked. A 24-volt tubular or strip heater may be installed inside the instrument to eliminate condensation. Alternatively, a number of silica-gel desiccators may be used, but these must be changed as soon as the crystals need regenerating.

2.5.4.2. *Alidade.* The angle of elevation is measured by means of an alidade. The standard alidade (see Plate XIII) consists of an engraved brass arc attached to a bracket which should be screwed to a post or corner of a building so that the plane of the engraved arc is accurately aligned with the searchlight beam. A movable pointer rotates about the centre of the arc and is provided with sighting points near each end. The observation is made by moving the pointer until the lower edge of the spot of light on the cloud base is in line with the two sights. The angle of elevation is then read from the engraved scale of degrees. It is essential of course that the line joining the two zero graduations on the scale should be truly horizontal.

2.5.5. Measurement of cloud height by balloon. Cloud height may be measured in daylight by determining the time taken by a small rubber balloon, inflated with hydrogen or helium, to rise from ground level to the base of the cloud. The base of the cloud should be taken as the point at which the balloon appears to enter a misty layer before finally disappearing.

The rate of ascent of the balloon is determined mainly by the free lift of the balloon and can be adjusted by controlling the amount of hydrogen or helium in the balloon. The time of travel between release of the balloon and its entry into the cloud is measured by means of a stop-watch. If the rate of ascent is n metres per minute and the time of travel is t minutes, the height of the cloud above ground is $n \times t$ metres, but this rule must not be slavishly followed. Eddies near the place of launching may prevent the balloon from rising until some time after it is released. Normally the stop-watch is started on the release of the balloon and therefore the elapsed time between release and the moment when the balloon is observed to have left the eddies will need to be subtracted from the total time before determining the cloud height. Even apart from eddy effects, the rate of ascent in the lowest 600 m (2000 ft) or so is very variable. Determinations of the height of very low clouds by pilot balloon are often appreciably in error on this account.

Although the height of base of medium cloud is sometimes obtained as a by-product in the measurement of upper winds by pilot balloon, the balloon method is mainly applicable to low clouds. Where no optical assistance is available in the form of binoculars, telescope or theodolite, the measurement should not be attempted if the cloud base is judged to be higher than about 900 m (3000 ft) unless the wind is very light. In strong winds the balloon may pass beyond the range of unaided vision before it enters the cloud.

Precipitation reduces the rate of ascent of a balloon and measurements of cloud height by pilot balloon should not be attempted in other than light

precipitation. The result of an ascent in light precipitation should be considered as a guide for estimating the probable height of the cloud base rather
than as a measurement.

The following instructions are addressed to observers at stations where
pilot-balloon equipment is used for the measurement of height of cloud.

2.5.5.1. *Equipment.* The standard equipment for the measurement of
height of cloud using hydrogen comprises:

Hydrogen cylinder with lever key	Balloon filler Mk 8
Pressure gauge and adaptor	Rawhide mallet
Rubber tubing	Stop-watch
Large adjustable spanner	Type CP10 balloons (10 gram, red)
Fine adjustment valve	Type CP30 balloons (30 gram,
Earthing equipment	orange).

2.5.5.2. *Hydrogen cylinders: safety precautions.* Being the lightest of
gases, hydrogen is very prone to leak from the containing cylinder. Hydrogen
is highly inflammable in air, and hydrogen–air mixtures are highly explosive
for a wide range of ratios of hydrogen to air; a small spark of low energy is
sufficient to cause ignition. Great care must be taken to comply with the
instructions given in 2.5.5.3 to 2.5.5.7.

2.5.5.3. *Storage of hydrogen cylinders.* Cylinders of hydrogen should, if
possible, be stored in the open near the filling shed. They should be covered
to protect them from the weather but they should be adequately ventilated.
Where exposure to the direct rays of the sun could result in a dangerous rise
in pressure (especially in tropical and subtropical regions) they should be
sheltered by a roof supported on four corner posts forming an open-sided
building.

If storage in the open is impracticable, a scheme for indoor storage should
be prepared in accordance with advice from Meteorological Office Headquarters.

2.5.5.4. *Care and use of hydrogen cylinders.* The presence of oil and
grease on cylinders containing gases can lead to serious explosions. As the
valve caps on hydrogen cylinders are interchangeable with those on a wide
variety of other cylinders, it is strictly forbidden to use lubricants to ease the
removal of caps from cylinders. The valve-protection cap should be fitted at
all times except when a cylinder is actually in use. The pressure of hydrogen in
a cylinder should be tested immediately the cylinder is received. The procedure is as follows:

(*a*) Unscrew the cylinder cap by hand. If the cap does not unscrew easily,
tap round the full circumference of the cap with a rawhide or similar
non-ferrous type of mallet. This tapping with the mallet must be confined to directly over the cap/collar threads. If the cap will not then
unscrew, the full cylinder should be returned to the supplier.

(*b*) Remove with a brush any dust or corrosion from the valve and orifice of
the cylinder. Wire brushes made of ferrous material must not be used
for this purpose.

(*c*) Fit the pressure-gauge adaptor to the main tap (left-hand thread), tightening it by hand or by using a rawhide or wooden mallet, not by using a

hammer or other object containing ferrous material. Then screw the pressure gauge into one end of the adaptor and the fine-adjustment valve into the other.

(*d*) Fit the key on to the cylinder cap and turn counter-clockwise not more than a quarter of a turn to release the hydrogen. The main valve of the cylinder should be opened or closed by hand; if this cannot be done, a rawhide or wooden mallet may be used to strike the valve key, but no hammer or object containing ferrous metal may be used.

(*e*) Note the pressure on the dial of the pressure gauge and then turn off the hydrogen, closing the cylinder valve tightly, and unscrew the pressure gauge and adaptor after making sure that the valve is not leaking. In the United Kingdom or other cool or temperate regions the pressure in a full cylinder should be about 120 atmospheres (or 1800 pounds-force per square inch (lbf/in^2)). If it is much greater than this, the excess should be allowed to escape gently, care being taken that the gas escapes to the open air and does not accumulate in an enclosed space. If the pressure in a new cylinder is less than 100 atmospheres (1500 lbf/in^2) a report should be made to the issuing authority.

For tropical or subtropical regions the pressure in a full cylinder should be only 100 atmospheres. Any excess pressure should be reduced, as indicated above, and any cylinders in which the pressure is less than 80 atmospheres (1200 lbf/in^2) are to be reported to the issuing authority.

Cylinders are not to be discharged below a pressure of 100 lbf/in^2. If the cylinder is completely emptied, air may get in and the water vapour would damage the inner lining.

2.5.5.5. *Balloon-filling shed.* The filling shed should, if possible, be a detached building having a sloping ceiling or roof with adequate ventilation from the highest point. The electrical wiring of buildings where hydrogen is handled should be external to the buildings and carried out in accordance with current regulations. Advice should be sought through Meteorological Office Headquarters before altering old buildings or erecting new ones.

A notice should be painted in bold red letters on the door of the shed:
EXPLOSIVE GAS
NO NAKED LIGHTS
NO SMOKING
No open flames, electric fires, running motors, etc. are permitted inside the shed, neither should combustible material be stored there.

The appropriate station authority must be informed of the use to which the building is to be put and invited to make any additional modifications which may be regarded as necessary by that authority.

2.5.5.6. *Earthing system.* When hydrogen is being discharged, the cylinders should be earthed, as should the balloon filler, except when the gas in the balloon is being adjusted for 'free lift'. A static charge of electricity is developed when hydrogen is being discharged or balloons are being filled and, especially in dry weather, the charge may build up sufficiently to cause a spark which could ignite the hydrogen.

Each hydrogen cylinder must be connected by a cable to an approved 'fixed earthing system'. A 'movable earthing system', supplied by the Meteorological Office (Met O 4, Western Road, Bracknell, Berkshire) must be connected to the balloon filler on each occasion a balloon is filled.

Each of the earthing systems must be examined by the particular designated authority at six-monthly intervals. In addition, a monthly check must be made at the station to see that all connections are firm and free from corrosion, holding screws are tight and the wire is in good condition. Specific instructions are issued to stations concerned.

2.5.5.7. *Inflation of balloons.*   To inflate Type CP10 (10 g) balloons:

(*a*) Fit the pressure-gauge adaptor to the cylinder main tap and screw down tightly (see 2.5.5.4(*c*) above).

(*b*) Fit the fine-adjustment valve to one end of the hexagon body of the adaptor and the pressure gauge to the other, screwing each up tightly. The tap to this fine-adjustment valve must be used to control only the flow of hydrogen—the cylinder key being used to turn the hydrogen on and off. The hydrogen pressure must not be allowed to fall below 100 lbf/in^2; when the pressure falls to this value the cylinder should be sent for recharging. New fine-adjustment valves incorporate a pressure-reduction valve.

(*c*) Clip the movable earthing system to the balloon filler and insert the valve of the filler into the 6 mm ($\frac{1}{4}$-inch) tubing connected to the fine-adjustment valve. The valve of the cylinder should then be opened momentarily to blow out particles of dust which might damage the balloon.

(*d*) Shake out any french chalk from the selected balloon, then fit the neck of the balloon over the nozzle of the filler. If the neck is rather large it may be made to fit more closely to the filler by rolling the neck back a turn or two.

(*e*) Turn on the flow of hydrogen with the cylinder key; then, by gently turning on the tap of the fine-adjustment valve, allow the hydrogen to fill the balloon. Fill until the balloon does not quite support the filler, then turn off the hydrogen with the cylinder key and examine the balloon for pin-holes. If any pin-holes are found, the balloon should be taken outside and the hydrogen allowed to escape through the neck, and the deflated balloon disposed of; on no account should it be released with the neck open. If no leaks are found, resume filling until the filler is raised just clear of the ground, then turn off the flow of hydrogen with the cylinder key and close the fine-adjustment valve gently until finger-tight.

(*f*) The operator must remove any electrostatic charge from his person before approaching the inflated balloon. This can be achieved by grasping an earthing point for a few seconds. The inflated balloon is not to be rubbed with the hand or allowed to rub against clothing. Remove the balloon and filler from the rubber tubing and unclip the movable earthing system from the filler. By means of the valve in the filler, release hydrogen until the balloon just supports the filler. Stand well away from the door of the filling shed to minimize the effects of draught when testing the balloon for balance and as a precaution against the balloon being carried away through the open door before the filler has been removed.

(g) Remove the filler, firmly pinching the neck of the balloon between the thumb and forefinger to ensure that hydrogen does not escape. Seal the balloon by stretching its neck and tying it firmly into a knot.

(h) Test for hydrogen leakage by placing a wetted finger over the end of the neck.

The balloon is now ready for release. If not required for immediate release it must not, under any circumstances, be allowed to ride against the ceiling; for parking purposes it is to be tethered to some object by a short length of string and is to be left free to move about in draughts without coming into contact with other objects.

When released, the balloon will rise at approximately 120 metres (400 ft) per minute. It should be released from a site which is free from obstacles in a downwind direction, i.e. the direction in which the balloon will travel.

Type CP30 (30 g) balloons are filled in the same way, but a special weight (weight B from the Mk 8 filler set) is screwed on to the filler; the balloon, when filled, will rise at approximately 150 metres (500 ft) per minute.

2.5.5.8. *Use of helium.* At stations supplied with helium for the filling of cloud-height balloons, the equipment supplied is the same as for use with hydrogen with the following exceptions. Because the threads on a helium cylinder are right-handed, an adaptor is supplied for screwing into the cylinder; the left-hand threaded fine-adjustment valve is then screwed into the adaptor. The earthing equipment is not supplied because helium, being an inert gas, does not form explosive mixtures.

The only precautions necessary are those which are normally taken with high-pressure cylinders: store in a cool place and do not drop the cylinder on a hard surface. Cylinders should never be discharged completely.

2.5.5.9. *Storage of balloons.* Balloons should be stored in a dark place at a temperature of between 5 and 25 °C. The balloons are usually wrapped in cellophane and packed in cartons which should not be opened until the contents are required for use. The cartons should be stored on shelves free from splinters and protruding nails. The shelves must not be near a source of heat or running electric motors. If the balloons are properly stored they should not deteriorate appreciably over a period of two or three years but bad storage conditions will ruin balloons in a period of weeks.

2.5.6. Cloud-base recorders. The Meteorological Office cloud-base recorders, Mk 3A and Mk 3B, illustrated diagrammatically in Figure 3, have been designed to aid observers in providing more accurate estimates of the heights of the cloud bases that they report. Observers should note that the cloud-base recorder can only record what is directly above the instrument and then only up to a limited height. Additionally, the reading of cloud base is influenced by the properties of the cloud and the setting of the receiver; the instrument is not an absolute device. The observer must consider the whole sky and only use the recorded information as a guide in arriving at the estimate of cloud

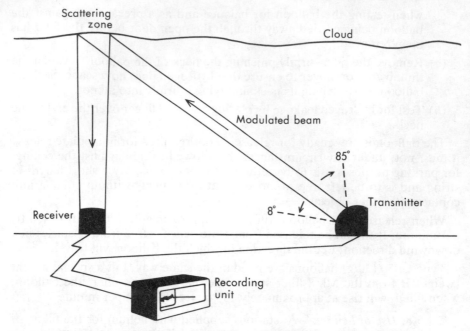

Figure 3.   Meteorological Office cloud-base recorder system

heights to be reported. The equipment is fully automatic and capable of unattended continuous operation both by day and by night.

The complete system consists of three units: a transmitter, a receiver and a recorder. The transmitter and receiver are remotely switched from the recorder which may be positioned away from the operating site by up to 1000 m for the Mk 3A and up to 24 km for the Mk 3B, although distance varies with cable impedance.

The transmitter emits a modulated light beam which sweeps in a vertical plane through elevation angles from 8° to 85°, and back to 8° once a minute. The receiver looks vertically upwards in the same plane and is sensitive only to light produced by the transmitter. In the recorder unit, a pen traverses a chart in step with the sweep of the transmitter, and is made to mark the chart whenever the receiver detects light from the transmitter beam which is reflected and scattered by cloud. The pen marks the chart during both upward and downward sweeps; when the base line from transmitter to receiver is 100 m (350 ft), the marks indicates light-scatter from cloud in the range 15 to 1200 m (50 to 4000 ft). More than one mark can be made during a single sweep. At the lowest point for each scan, the beam is deflected along a horizontal path and reflected from a bar into the receiver. This signal produces a zero mark for each scan on the chart.

The chart is printed with a height scale and it is the lowest value of each pen mark that represents the response to cloud base. Plate XIV illustrates two examples of cloud-base recorder charts.

These recorders will eventually be superseded by a cloud-base recorder using a laser technique.

CHAPTER 3

VISIBILITY

3.1. INTRODUCTION

Visibility is defined as the greatest distance at which an object can be seen and recognized in daylight, or at night could be seen and recognized if the general illumination were raised to a daylight level. The criterion of recognizing an object must be used and not merely the seeing of an object without recognizing what it is. For meteorological purposes it is necessary that visibility observations give a measure of the transparency of the atmosphere. Other factors, however, affect the range at which an object can be seen (e.g. its size, colour, background, etc.—see 3.2.1). By selection of appropriate objects and when observing in suitable conditions the effect of these extraneous factors can usually be eliminated in daylight and a meteorological visibility dependent only on the optical state of the atmosphere can be observed. The visibility at night, as defined above, cannot of course be observed. In practice, lights may be used instead of objects, but the range at which lights can be seen depends not only on the atmospheric transparency but also on other factors such as the luminous intensity of the light source and the illumination of the background (see 3.4). A relationship must be established between the visual range of the lights at night and the equivalent daylight meteorological visibility.

In order to improve the scientific basis of visibility measurements, the World Meteorological Organization (WMO) recommended in 1957 the adoption of a new measure of the optical state of the atmosphere—the meteorological optical range (MOR). This is defined as the length of path in the atmosphere to reduce the luminous flux in a collimated beam from an incandescent lamp at a colour temperature of 2700 K to 0·05 of its original value. This unit depends solely on the atmospheric transparency along the path. The MOR can be measured accurately with suitable instrumentation both by day and by night.

The visual extinction coefficient, σ, is defined as the fractional reduction in luminous flux (Φ) of a collimated beam of monochromatic light in unit distance (s), i.e.

$$\sigma = -\frac{1}{\Phi}\frac{d\Phi}{ds}.$$

Integrating,

$$\frac{\Phi}{\Phi_0} = e^{-\sigma s},$$

where the luminous flux is reduced from Φ_0 to Φ in distance s. Thus the MOR (M), by definition, can be obtained from

$$\frac{\Phi}{\Phi_0} = 0{\cdot}05 = e^{-\sigma M},$$

$$\text{or } M = \frac{-\log_e 0{\cdot}05}{\sigma}. \qquad \qquad \dots (1)$$

43

Koschmieder (as quoted in Middleton*) has shown that the meteorological visibility in daylight (V) is related to σ by the equation

$$V = \frac{-\log_e \epsilon}{\sigma}, \qquad \qquad \ldots (2)$$

where ϵ is the threshold of contrast between the object and its background to allow the object just to be recognized. The value of ϵ is believed to be about 0·05; this value gives equality between M and V. Thus, if σ is measured from observations of lights at night, the MOR which is equivalent to the daylight visibility can be approximately obtained from equation (1).

Although meteorologists are primarily interested in the MOR and their observations of visibility are aimed at measuring this, the term 'visibility' is still used almost universally in lieu of MOR and this practice will be followed in this handbook. The lower ranges of visibility are expressed in metres (m) and the higher ranges in kilometres (km).

The practical application of the definition is dealt with in subsequent paragraphs. The advice presupposes an observer with normal eyesight, observing without the aid of binoculars, telescope or theodolite.

3.2. VISIBILITY OBJECTS

The assessment of the visibility in daylight is generally based on the observation of suitable objects at known distances.

3.2.1. Suitability of objects for visibility observations. The principles involved in the selection of the objects are (a) the objects should be black or very dark coloured and stand above the horizon when viewed from the normal observing point, and (b) they should subtend an angle of at least 0·5° in width and elevation to the observer but not more than 5° in width. This upper limit can easily be exceeded with near objects, unless care is taken in selection.

If an object with a terrestrial background has to be selected, it should stand well in front of the background so that the distance between the object and background is at least half that of the object from the observing point. A tree at the edge of a wood, for example, would not be suitable for visibility observations. A white house would be an unsuitable object, particularly when the sun is shining on it, but a group of dark trees would be satisfactory except when brightly illuminated by sunlight.

In order that visibility measurements should be representative and comparable on an international scale, the measurements should refer to objects large enough to subtend an angle of 0·5° at the observer. This angle may be estimated by making a hole 7·5 mm in diameter in a card. When the card is at arm's length such a hole will subtend the specified angle at the observer's eye. The visibility object viewed through such an aperture should completely fill it. As a guide to size, an angle of rather more than half a degree is subtended by an object 1 m wide at 100 m, 10 m wide at 1000 m, and so on in proportion.

---

*Middleton, W. E. K. *Vision through the atmosphere*. Toronto, University of Toronto Press, 1952.

Thus, suitable objects need to be of the dimensions of a bush at 100 m, a house at 1000 m, a church at 2 km, and a hill over 150 m high at 15 km. In practice, less bulky objects may have to be selected because objects complying with the specifications are not available. At distances beyond 1000 m where there is a choice of objects to select, the larger should be selected. For distances beyond a few kilometres it is usually necessary to select topographical features such as ridges or hill tops, though these are not ideal objects because they may be hidden at times by a low cloud, or their prominence modified by seasonal features such as a covering of snow.

3.2.2. Selection of visibility objects at synoptic stations. Observations of visibility at synoptic stations, and especially at aerodromes, are of great importance. A good observer should be able to attain an accuracy of reporting to within 10 per cent of the actual visibility, and visibility objects should be carefully and comprehensively selected with this in mind. All suitable objects should be used and, where possible, two or more objects at the same distance but in different directions should be selected as this will enable the observer to recognize differences of visibility in various directions. This latter point is especially important at coastal stations where islands, light-towers, fixed buoys and the distance of the horizon from the height of the observer's eye may all be used to assess visibility across the sea.

3.2.3. Selection of visibility objects at climatological stations. The visibility at a climatological station is not required to the same degree of accuracy as that necessary at a synoptic station. Visibility is recorded as being within one of a series of visibility ranges, and visibility objects should, whenever possible, be chosen at the standard distances given in the table below.

Object	Standard distance	Permissible variations	General description of visibility	Coded entry on monthly return from climatological station
	m	*m*		
—	—	—	Dense fog	X*
A	20	18–22	Dense fog	F
B	40	36–44	Thick fog	0
C	100	90–110	Thick fog	1
D	200	180–220	Fog	2
E	400	360–440	Moderate fog	3
F	1000	900–1100	Very poor	4
	km	*km*		
G	2	1·8–2·2	Poor	5
H	4	3·6–4·4	Moderate }	6
I	7	6·3–7·7	Moderate }	6
J	10	9–11	Good	7
K	20	18–22	Very good }	8
L	30	27–33	Very good }	8
M	40	36–44	Excellent	9

*Visibility less than 20 m.

3.2.4. Distance of visibility objects. The distance of selected objects should be carefully determined, preferably by direct measurement for the nearer objects and calculation from a large-scale Ordnance Survey map for the more distant ones.

3.2.5. Record of visibility objects.   The selected visibility objects should be listed at climatological stations in Metform 3100A (Supplement to the Pocket Register), and at synoptic reporting stations in the appropriate pages of the Register. A copy of the list should be displayed conveniently for the observer, together with a key map indicating the objects in a diagram or panoramic photographs, and including bearings from the observing point in degrees from true north.

3.3. DETERMINATION OF VISIBILITY DURING DAYLIGHT

3.3.1. General principles.   The assessment of the visibility during daylight is generally based on the observation and recognition of suitable objects at known distances. Assuming that the observer has normal eyesight, the distance at which objects of suitable size, as explained in 3.2.1, can be seen depends mainly on such factors as:

(*a*) the transparency of the atmosphere,

(*b*) the position of the sun, and

(*c*) the degree of contrast between an object and the background.

As the visibility should depend only on factor (*a*), it is necessary to take precautions to eliminate the effects of (*b*) and (*c*). Thus if the sun is shining it is desirable to observe objects located at an angle of 90° or more from the sun's direction and, in particular, to avoid viewing an object against a rising or setting sun as this may suggest an exaggerated visibility. Whenever possible, the objects used should be those visible against the sky, or having a good background contrast.

If possible, the observation should be made from a position where the observer has an uninterrupted view of the entire horizon. If this is not possible the observer should change viewpoints until he has viewed the horizon in all directions. As the requirement is for horizontal visibility at the earth's surface, the eye of the observer should be at normal height above the ground; thus measurements should not be made from high buildings.

3.3.2. Observations of visibility.   If one of the visibility objects is just recognizable for what it is known to be, then its distance is the visibility. A perfect set of visibility objects (each within 10 per cent of the distance of the next object at synoptic stations, or at each of the standard distances at climatological stations) is rarely, if ever, available. In practice, therefore, when one object is seen with clarity but the next furthest object is either not visible or is blurred or indistinct, then the visibility is somewhere between the distances of the two objects; in these circumstances an estimate of the visibility must be made of the distance at which an imaginary suitable object could be seen and identified, based on the relative clarity and distance of the two known objects.

Visibilities greater than the distance for which objects are available may be estimated from the general transparency of the atmosphere. This can be done by noting the clarity with which the furthest visibility object stands out. Sharp outlines and relief, with little or no blurring of colours, indicates that the visibility is much greater than the distance of the object. If the furthest visibility object, say at 15 km, is seen with the clarity as described above, an experienced observer may estimate the visibility to be 40 km; at places where the horizon is restricted estimations of this kind may frequently have to be made.

Whereas at synoptic stations the aim should always be to observe visibility to an accuracy within 10 per cent, at climatological stations it is necessary merely to record the correct visibility letter as given in the table on page 45. For example, if an object at distance D can be seen, but not the object at distance E, then the correct entry is D. Where suitable objects are not available at all the standard distances, letters should be selected by estimation as indicated above. If the object at distance A cannot be seen, the entry is X.

3.3.3. Visibility varying in different directions. When the visibility varies markedly in different directions the lowest visibility is logged. A note should be made in the remarks column detailing the maximum visibility and the directions from the observer at which this occurs; similarly the direction of the minimum visibility should be recorded together with a brief comment, if appropriate, giving an explanation for the disparity. At stations which make full synoptic reports the maximum and minimum visibility (when the visibility is less than 1000 m) and the associated directions from the observer may be added to reports in the form described in the *Handbook of weather messages*, Parts II and III.

3.3.4. Coastal stations. The visibility to be logged is that over land. If the visibility over the sea is different from that over the land, the seaward visibility should be noted in the remarks column and at full synoptic stations may be added to transmitted reports (as detailed in the *Handbook of weather messages*, Parts II and III) if the station is instructed to do so.

3.4. DETERMINATION OF VISIBILITY AT NIGHT

Visibility reports at night should indicate the same degree of atmospheric transparency as they do by day. The change from daylight to darkness does not itself affect the visibility; if changes do occur they are the result of an alteration in atmospheric conditions.

The most suitable objects for determining the visibility at night are unfocused lights of moderate intensity at known distances and the silhouettes of hills and mountains against the sky.

The list of visibility objects at a station should also include lights suitable for determining the visibility at night.

The distance at which a light can be seen at night depends mainly on the following factors:

(a) luminous intensity of the light,

(b) sensitivity of the observer's eyes,

(c) presence or absence of other bright lights in the field of view,

(d) general level of illumination, and

(*e*) transparency of the atmosphere.

In determining the visibility from the observation of fixed lights the effects of the first four factors should be reduced as far as possible. In order that the observer's eyes should become as well adjusted as possible, allow at least two minutes in darkness for the eyes to adapt to night vision.

The practical ways of determining the visibility from fixed lights includes the use of the Meteorological Office visibility meter Mk 2 (formerly called the Gold visibility meter) and the use of lights of known luminous intensity (described in section 3.4.3).

3.4.1. Instructions for the use of the Meteorological Office visibility meter Mk 2.

3.4.1.1. *Description of the visibility meter.*   The visibility meter (Plate XV) is a simple visual photometer used to measure the luminous flux from a distant light, and hence the transparency of the atmosphere. This is done by observing the lamp through graduated neutral filters which can be varied until the lamp is only just visible. The main filter is about 20 cm long and varies in opacity in a regular manner along its length, being almost transparent at one end. It slides in a frame to which are fixed two small filters (about 2 cm square) whose opacity varies in the same manner as that of the main filter, but in the opposite direction. One of the small filters is almost transparent while the other transmits only about 1/1000 of the incident light. The effect of superimposing either of the two small filters on any part of the main filter gives an area of uniform opacity, which is continuously variable by sliding the main filter. The opacity is measured by two scales, corresponding to the small filters, on either side of the main filter, which move past two fixed marks on the frame. The unit employed is the 'nebule', which is defined by the statement that a filter of opacity 100 nebules transmits 1/1000 of the incident light, which implies that a screen of opacity 1 nebule has a transmittance of 0·933. The scales are linear, running from 15 to 120 nebules for the clearer of the small filters, and 115 to 220 nebules for the denser one.

3.4.1.2. *Installation of visibility lights.*   Fixed lights of constant intensity must be used, and it is most convenient to have at least three lights arranged so that there is one at approximately 100, 450 and 1350 m from the observer. The lights should be 2 to 3 m above the ground, and should have some form of hood to protect them from the weather (and from theft), but which is removable to facilitate cleaning and replacing of the lamps. The hood should be black inside and no lens or focusing mirror should be used. A 15-watt lamp should preferably be used at 100 m, a 100-watt at 450 m and a 100-watt (or higher if possible) at 1350 m.

On aerodromes the light should be well screened so as to be visible only from near the observation point, and should be sited so as not to shine along any runway; they must not cause an obstruction to the runway or perimeter track. Approval must be obtained from the aviation authorities before the lights are sited.

It is not essential that the lamps should be switched on and off from near the observation point; it may often be much cheaper to install a switch on the light itself, leaving it on all night. Alternatively the lights could be wired into a road-lighting or obstruction-light circuit, provided it is (or can be) in use all night. Circuits subject to large voltage variations (either deliberate or accidental) must be avoided.

Where the lights are on for long periods, the lamps must be renewed at regular intervals, e.g. monthly in winter and quarterly in summer. The rated life of a domestic tungsten-filament lamp is normally 1000 hours. A tungsten-filament lamp will emit an approximately steady light after having been 'aged' by being continuously lit at its rated voltage for a period of 5 per cent of its rated life, that is normally for 50 hours. It is desirable that the new lamp should be of exactly the same type as the old. Each observer should make new calibration figures every time a lamp is replaced, lamps being aged at the station before use.

The observations are greatly facilitated if the lights are observed through fixed tubes, one for each light. The tubes may be lengths of iron pipe, painted matt black inside, about 1 m long and 3·5 cm in diameter. It is preferable to fix them through the wall of a room, if a suitable room is available, that is one from which all the lights can be seen, and which can be darkened for two minutes before each observation. The advantage of this arrangement are threefold:

(*a*) There is never doubt about which light is being observed, as each tube points directly to its own light.

(*b*) The observer can be in complete darkness, and shielded against any extraneous lights.

(*c*) The observations can be made in greater comfort, especially in cold or wet weather, and will therefore be more reliable.

When the observation must be made out of doors, with or without the assistance of tubes, the observing point should be in a position well away from any artificial light.

When the installation is complete, the distance of each light from the observation point should be carefully measured.

3.4.1.3. *Preparation of conversion diagrams.* These are used to obtain the equivalent daylight visibility from the meter readings. A card (Metform 2812, Conversion diagrams for use with visibility meter) contains skeleton diagrams and full instructions for their completion. If this card is not available, diagrams can be prepared locally (one for each light) by marking along a straight line two scales, one of nebules and one of visibility in metres. The two scales must conform with the relation:

$$\text{number of nebules} = 43 \times \frac{\text{distance of light (metres)}}{\text{visibility (metres)}}.$$

The scale should not be extended below 10 nebules unless this is necessary to provide an overlap with the next more distant light. Part of a specimen set of scales applicable to a light at exactly 100 metres is reproduced below.

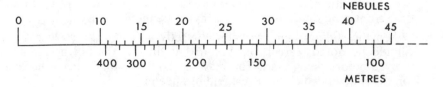

3.4.1.4. *Individual calibration figures.*   Before the meter can be brought
into routine use, each observer must determine his calibration figure for each
light; this is the meter reading which would be obtained if the atmosphere
were perfectly transparent. In practice this should be done immediately after
dark on an evening when the visibility has been observed (during daylight) to
be not less than 12 times the distance of the light in question from the observer and is not expected to change much, for example when there is a cloudy
sky, a fresh wind, and humidity not over 95 per cent. The procedure described in the next section should be carefully followed, and not more than
one light should be observed at one time, as the longer time taken would
improve the adaptation of the eyes to the dark and thus cause non-
representative readings.

An observer, having taken a meter reading M, obtains from it his personal
calibration figure for the light in question by calculating the following expression:

$$\text{calibration figure} = M + 0 \cdot 043 \times \frac{\text{distance of light (metres)}}{\text{visibility (kilometres)}}.$$

A calibration book should be kept, with a page for each observer showing
his calibration figure for each light and with the date when it was determined.
New figures should be determined on every convenient occasion, and comparison with earlier figures will give the best average figure for future use. A
recalibration should always be made whenever a lamp is changed.

3.4.1.5. *Routine use of the visibility meter.*   The procedure for making an
observation with the meter is as follows:

(*a*) Ensure that the glass surfaces of the meter are clean. If necessary, clean
     them with a lint-free cloth. (When not in use the meter and the cloth
     should be kept in a closed box.)

(*b*) Allow about two minutes for the eyes to adapt to the darkness before
     beginning to make an observation.

(*c*) Observe the most distant light which is visible through the visibility
     meter, and adjust the sliding filter until the light is only just visible. This
     should be done by pulling out the slide until the light just disappears,
     then pushing it in slowly until the light just reappears. Always begin
     with the eyeshield over the clearer small filter, but if the light is still
     visible with the slide pulled right out, the eyeshield should be transfer-
     red to the dark filter. Look directly at the light during the observation
     as the sensitivity of the eye is more constant for direct vision. Do not
     spend a long time making an observation as the eyes slowly adapt to the
     dark and this will invalidate the current calibration figure for an indi-
     vidual. The observer should aim at always spending three minutes be-
     tween leaving a well-lighted room and completing the observation. If
     the observer normally wears spectacles for distant vision he should
     wear them for making observations, but they must be kept clean and
     free from condensation or rain (as must the meter itself).

(*d*) When the sliding filter has been correctly adjusted the scale reading
     should be noted, care being taken that the slide is not accidentally
     moved before being read. The reading should be taken from the scale
     on the slide on which the eyeshield has been used.

(*e*) This reading should be subtracted from the observer's calibration figure for that particular light; the resulting figure is then used, in conjunction with the appropriate conversion diagram, to read off the visibility.

(*f*) If the difference between the calibration figure and the observed reading is less than 10 nebules, even for the most distant light, the diagrams cannot be used. The visibility is then known to be over 4·3 times the distance of the furthest light and it must be estimated visually. It will be evident that this visibility meter requires more care in its use than the majority of meteorological instruments. Lack of care, for instance in neglecting to allow always the same length of time for the eyes to adapt to the dark, will lead to considerable errors.

3.4.2. Transmissometer. The transmissometer measures the luminous flux of a beam of light after it has passed a known distance through the atmosphere. By comparing the luminous flux with that which would be obtained in perfectly clear conditions, it is possible to calculate the extinction coefficient of the air, and hence the visibility (see equation (1) in 3.1). The design of the transmitter and receiver housings limits the field of view, and hence the effect of background lighting, and this allows the transmissometer to operate both during day and night. The main advantages of the transmissometer over the visibility meter are the independence of sensitivity of the observer's eye and the ability to operate throughout the full 24-hour period.

In the Meteorological Office transmissometer the source of light is a low-voltage projector lamp. The luminous flux of the beam of light from this source, after it has passed through a 200-metre horizontal path of the atmosphere at a height of 3 metres, is measured by means of a photo-electric cell housed in the receiver unit. A means of adjusting the receiver gain is incorporated for calibration purposes. The lamp runs from a stabilized mains supply through a step-down transformer. The range of a transmissometer is related to the baseline, and for the 200-metre baseline the operating range is about 100 metres to 10 kilometres. Within that range the transmissometer will measure visibility to within 11 per cent accuracy whilst giving an indication, at a lower accuracy, outside that range.

The differences between the Mk 3A, Mk 3B and Mk 4 versions are only in the telemetry of data and the processing of the signal. Detailed operating instructions are available from the Operational Instrumentation Branch.

Observers should note that both the visibility meter and transmissometer measure visibility in only one direction and, moreover, over a restricted distance. When using these instruments a visual check should always be made in other directions.

3.4.3. Auxiliary estimates of visibility at night by direct observation of lights. If the visibility meter cannot be used, less precise estimates of visibility can be made by eye observations of lights. To obtain consistent results, all suitable lights should be listed, giving the visibility limits (obtained as described in 3.4.3.3) below which each light becomes invisible. This should be done even at stations where the visibility meter is used, as the experience gained will be of great assistance if the meter becomes unserviceable and is especially useful in determining visibility in different directions.

3.4.3.1. *Luminous intensity and colour of selected lights.* Each selected light must be of constant and known rating and luminous intensity (candela) which should not vary greatly with the direction from which is is viewed: lights fitted with lenses or mirrors to throw the light in certain directions rather than others should not be used. Flashing lights may be used provided each flash lasts for one second or more. The luminous intensity (in candelas) should be ascertained from the engineer or other authority in charge of the lights, or from the manufacturers. Ordinary domestic electric-light bulbs of 100, 60 and 15 watts have luminous intensities of approximately 92, 46 and 9 candelas, respectively. On aerodromes, red obstruction lights on low structures may also be used but, because of the absorptive properties of the red glass, the luminous intensity will then be very much less than that of the bulbs themselves. The following table may be used for obtaining the luminous intensity of standard types of red obstruction lights:

No. of bulbs in the light	Power per bulb (watts)	Luminous intensity (candelas)
4	75	22
2	75	15
2	60	10
2	15	3

Possible unserviceability of one or more of the bulbs and dirt or ice on the glass covers would reduce the above values of luminous intensity emitted, and the resulting estimate of visibility would be below the actual visibility.

Alternatively the luminous intensity of the light can be measured with the visibility meter on a night of good visibility (15 km or more). The observer should stand at a carefully measured distance from the light, between 10 and 100 metres according to its intensity, and should adjust the visibility meter until the light is only just visible through it. This should be completed three minutes after leaving a lighted room. If the meter reading is N (nebules) and the distance from the light is D (metres), the intensity I (candelas) is given by:

$$\log I = 0{\cdot}03N + 2 \log D - k$$

where the value of k should be taken as

6 during twilight
6·7 in moonlight
7·5 in complete darkness.

The selected lights should preferably be white.

3.4.3.2. *Visual threshold.* The sensitivity of the eye is expressed in terms of the visual threshold: the illumination produced at the eye by a light so faint that it can only just be seen. It varies somewhat from one observer to another and for the same observer at different times. For practical purposes average values have been used in preparing Figures 4(a), (b) and (c) corresponding to (a) twilight or when there is appreciable light from artificial sources, (b) moonlight or when it is not quite dark, and (c) complete darkness or with no light other than starlight.

Except for red light, the sensitivity of the eye for indirect vision (looking a little to one side of the light) is greater than for direct vision after a few minutes have been spent in the dark. The eye's sensitivity for indirect vision (again except for red light) continues to increase for an hour or more after the observer has gone from a lighted room into weak illumination or darkness, whereas dark adaption for direct vision is complete in about two minutes. Indirect vision should therefore not be used for visibility observations, and observers must make sure that for this purpose they regard a light as visible only if they can see it when looking directly at it.

3.4.3.3. *Use of nomograms to determine visibility.* The relation between daylight visibility and the distance at which a light is just visible at night is given in Figures 4(a), (b) and (c). Each nomogram shows the distances at which lights of varying luminous intensity (from 1 to 1 000 000 candelas) are just visible when the equivalent daylight visibility has any value between 10 metres and 10 kilometres. The horizontal line corresponding to the distance of the light from the observer should be followed until it meets the curve corresponding to its luminous intensity (interpolating as necessary). The abscissa of this point gives the equivalent daylight visibility, and this should be inserted in the list referred to in 3.4.3.

It is essential that the appropriate nomogram is used, and three separate lists made for the specified conditions.

When very powerful lights are observed in poor visibility they may themselves make the background so bright as to raise the observer's visual

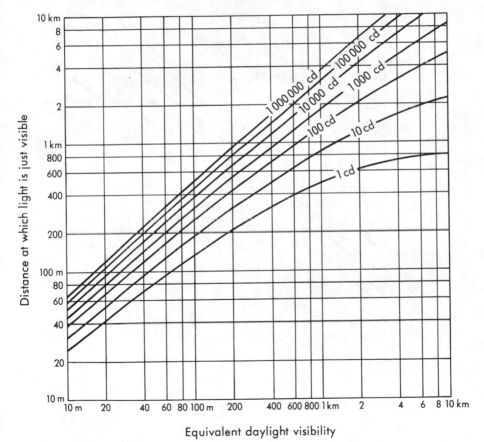

Figure 4(a). Twilight nomogram of visibility of lights at night

threshold. This effect is too variable to be readily allowed for, and the use of such lights for visibility observation should be avoided.

Figures 4(a), (b) and (c) are based on the equation of diminution of light in the atmosphere

$$E_t = \frac{I}{L^2}e^{-\sigma L}$$

and Koschmieder's formula (given earlier in equation (2) in 3.1)

$$V = \frac{-\log_e \epsilon}{\sigma},$$

where

I = luminous intensity of light (candelas),
L = distance at which light can just be seen at night (metres), and
E_t = threshold of illumination of the observer's eye for point-light sources at night (lux).

It is assumed that $E_t = 10^{-6}$, $10^{-6.7}$ and $10^{-7.5}$ lux, respectively, for twilight, moonlight and darkness, and $\epsilon = 0.05$. These values are recommended by WMO.

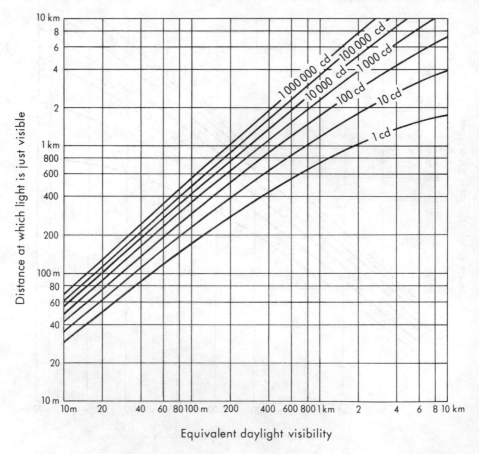

Figure 4(b).   Moonlight nomogram of visibility of lights at night

It is easily seen from Figures 4(a), (b) and (c) how misleading visibility observations can be if they are based simply on the distance at which ordinary lights are visible, without due allowance for the intensity of the light.

3.5. VERTICAL VISIBILITY

When the sky is obscured because of fog, falling or blowing snow or some other obscuring phenomena, observers are required to measure or estimate the vertical visibility in place of cloud height.

The vertical visibility is defined by the World Meteorological Organization as the vertical visual range into an obscuring medium. In the United Kingdom this definition is interpreted as the visual range of a dark object of moderate size viewed vertically upwards against a sky background in daylight.

Vertical visibility, as defined for United Kingdom use, cannot be estimated in darkness, nor can it be measured by either the cloud searchlight or cloud base recorder, because neither of these instruments measure visual range.

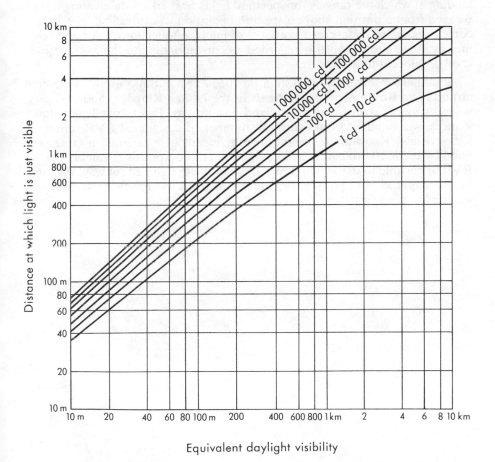

Figure 4(c). Darkness nomogram of visibility of lights at night

The vertical visibility may be taken as the height above ground at which a pilot balloon, *if rising vertically,* would disappear from view in daylight.

When the obscuring medium is so dense that the tops of tall structures are invisible from a point near their base, an estimate of vertical visibility can be made if the observer knows the height above ground of various points on the structures. An open structure, such as a lattice mast, is preferable to a solid building.

If the observer has no basis for making an estimate which he considers reasonably reliable he should omit the observation. The method of reporting vertical visibility is described in the *Handbook of weather messages,* Part II, and *Abbreviated weather reports.*

3.6. RUNWAY VISUAL RANGE

Runway visual range (RVR) is defined by the International Civil Aviation Organization (ICAO) as 'the maximum distance in the direction of take-off or landing at which the runway, or specified lights or markers delineating it, can be seen from a position above a specified point on its centre line, at a height corresponding to the average eye-level of pilots at touchdown'. A height of 5 metres above ground level is regarded as corresponding to the average eye-level of pilots at touchdown.

A 'human observer' RVR system has been brought into use at some civil airports and Royal Air Force airfields in the United Kingdom and overseas, whereby an observer (at a known fixed position, as close to the Touch-down Zone as safety allows) can count the number of designated RVR reference lights visible. Knowing this count, and by means of a calibration table supplied by the Meteorological Office, the Air Traffic Controller can obtain the RVR applicable to the Touch-down Zone of the designated runway. At some of the larger civil airports an instrumental RVR system using transmissometers has been installed.

CHAPTER 4

WEATHER

4.1. NATURE OF THE OBSERVATIONS

The observations to be recorded under the headings 'present weather' and 'past weather' include the following phenomena:

(a) Precipitation (including 'showery' precipitation) comprising rain or freezing rain, drizzle or freezing drizzle, sleet,* snow, snow pellets, snow grains, ice pellets, hail, small hail, and certain fine ice crystals referred to as 'diamond dust'.

(b) Atmospheric obscurity and suspensoids comprising haze, mist, smoke, fog or ice fog, drifting or blowing snow, duststorms and sandstorms, and dust whirls or sand whirls.

(c) Spout.

(d) Squalls.

(e) Lightning, thunder and thunderstorms.

(f) State of the sky.

These phenomena, termed 'meteors' (see 4.2) with the exception of (d) and (f), form the basis of the international weather code which is used in compiling synoptic reports, as detailed in the *Handbook of weather messages,* Part II. In climatological records their occurrence in present and past weather is noted in Beaufort letters (see 4.5). In synoptic reports their occurrence in past weather is noted in Registers in Beaufort letters and a code figure. The record of past weather should be as complete as possible even though all the information cannot be included in the coded report. For present weather, only the code figures are noted. It is essential that a more or less continuous watch should be kept so that the weather between observations can be accurately recorded. At synoptic stations, in particular, any sudden or significant changes which may occur should be recorded with the least delay.

4.1.1. Climatological stations. The observer enters in the Register the code figure for the weather at the time of the observation. Additionally, he is asked to note the times of beginning and ending of precipitation, fog, thunderstorms and other phenomena so that a brief account of the weather of the whole day may be entered in the weather diary on the climatological return. Details of the present-weather code used at climatological stations, and Beaufort letters and symbols for use in the weather diary, are given in Metform 3100A (the supplement to the Pocket Register for climatological observations.).

4.1.2. Synoptic stations. Present weather is entered as code figures and the observer also records in the appropriate column of the Daily Register a

---

*The term 'sleet' is commonly used in the United Kingdom to describe precipitation of snow and rain (or drizzle) together, or of snow melting as it falls, but it has no agreed international meaning.

Beaufort letter, or groups of Beaufort letters (see 4.5), describing as exactly as possible the weather since the last observation. The observer amplifies this information, whenever possible, by noting in the remarks column of the Register the time of beginning and ending of precipitation, fog, thunderstorms etc. to maintain an exact and detailed record of the weather. Provision is made in Beaufort letters for indicating the intensity of many phenomena, and the value of the record is increased by adding as much amplifying detail of this kind as possible. The reports of present weather and past weather have to be made in accordance with the synoptic codes as detailed in the *Handbook of weather messages,* Parts II and III, and *Abbreviated weather reports.*

4.2. DESCRIPTIONS OF PHENOMENA

Most of the phenomena recorded in weather observations are composed of meteors. The *International cloud atlas,* Volume I, 1975 gives the following definition:

> A meteor is a phenomenon observed in the atmosphere or on the surface of the earth, which consists of a suspension, a precipitation, or a deposit of aqueous or non-aqueous liquid or solid particles, or a phenomenon of the nature of an optical or electrical manifestation.

Meteors are classified in four groups: hydrometeors (see 4.2.1), lithometeors (see 4.2.2), electrometeors (see 4.2.3) and photometeors (see Chapter 11).

Photometeors (luminous phenomena such as rainbows) are not items of weather as the term is generally understood, but their occurrence should be recorded because of their general meteorological interest.

4.2.1. Hydrometeors.   A hydrometeor is a meteor consisting of an aggregate of liquid or solid water particles

(a) falling through the atmosphere (rain, drizzle, snow, snow pellets, snow grains, ice pellets, hail, small hail and diamond dust); they originate mostly in clouds and commonly reach the earth's surface (precipitation), though they may completely evaporate during the fall (virga);

(b) suspended in the atmosphere (clouds, fog, ice fog, mist);

(c) blown by the wind from the earth's surface (drifting snow, blowing snow and spray) and generally confined to the lowest layers of the atmosphere; or

(d) deposited on objects on the ground or in the free air (dew, hoar frost, rime, glaze, fog droplets) occurring either in the form of aggregates of particles, more or less individually discernible in spite of the fact that they are often partially linked together (hoar frost, rime), or as smooth homogeneous layers in which no pellet structure can be distinguished (glaze).

The most common hydrometeors are listed below with a brief description of each.

4.2.1.1.   *Rain:*   precipitation of drops of water (by convention having a diameter of more than 0·5 mm) from a cloud. The diameter and concentration of drops vary considerably according to the intensity of the precipitation and especially according to its nature and source (continuous rain, rain showers, etc.).

Freezing rain: raindrops with the temperature below 0 °C and which freeze on impact with the ground or with objects on the earth's surface. (It is of course assumed that the objects are not artificially heated above, or cooled below, the temperature of the surrounding air.)

4.2.1.2. *Drizzle:* fairly uniform precipitation comprised exclusively of very fine drops of water (less than 0·5 mm in diameter) and very close to one another. The drops appear almost to float, thus making visible even slight movements of the air, and the effect of their individual impact on water surfaces is imperceptible.

Drizzle falls from a fairly continuous and dense layer of stratus cloud, usually low, sometimes touching the ground (fog).

Freezing drizzle: drizzle-drops with temperature below 0 °C and which freeze on impact with the ground or with objects on the earth's surface. (It is again assumed that the objects are not artificially heated above, or cooled below, the temperature of the surrounding air.)

4.2.1.3. *Snow:* precipitation of ice crystals, most of which are branched, from a cloud. The form, size and concentration of snow crystals differ considerably according to the temperature at which they form and the conditions in which they develop. At temperatures warmer than about −5 °C the ice crystals are generally agglomerated into snowflakes. Small flakes, up to 4 or 5 mm in diameter, especially those occurring at the beginning of a snowfall in very cold weather, often show a six-rayed starlike structure of great beauty. Larger flakes usually consist of tangled aggregates of such crystals so that the geometrical structure ceases to be perceptible.

4.2.1.4. *Snow pellets:* precipitation of white and opaque ice particles which are generally rounded but sometimes conical; their diameter is in the range 2–5 mm.

The pellets are brittle and easily crushed; when they fall on hard ground they bounce and readily break up. Precipitation of snow pellets generally occurs in showers, together with precipitation of snowflakes or raindrops, when surface temperatures are around 0 °C.

4.2.1.5. *Snow grains:* precipitation of very small, white, opaque particles of ice which are fairly flat or elongated; their diameter is generally less than 1 mm.

When the grains hit hard ground they do not bounce or shatter. Except in mountainous areas, they usually fall in small quantities, mostly from stratus or from fog, and never in the form of a shower. This precipitation corresponds as it were to drizzle, and occurs when the temperature is in the approximate range of 0 °C to −10 °C.

4.2.1.6. *Ice pellets:* precipitation of transparent ice particles which are spherical or irregular, rarely conical, and which have a diameter of less than 5 mm. Usually ice pellets are not easily crushable; when they fall on hard ground they generally bounce with an audible sound on impact. Precipitation in the form of ice pellets generally falls from altostratus or nimbostratus.

4.2.1.7. *Hail:*   precipitation in the form of either transparent, or partly or completely opaque, particles of ice (hailstones). They can be spheroidal, conical or irregular in form, with a diameter between about 5 and 50 mm. They fall, either separately or agglomerated into irregular lumps, only in showers and are generally observed during heavy thunderstorms. Hailstone structure varies from alternate layers of opaque and transparent ice (usually the more common variety) to only transparent or opaque ice which has formed around a core which is not necessarily at the geometric centre.

If the opportunity arises and the facilities are available, large hailstones should be weighed and measured and, if possible, photographed whole and in cross-section. Failing this, comments on their structure and size, determined before they disperse, can be of value.

4.2.1.8. *Small hail* (formerly called 'ice pellets, type (b)'):   precipitation of translucent ice particles, almost always spherical but sometimes having conical tips. Their diameter may attain and even exceed 5 mm.

Small hail consists of snow pellets in a thin layer of ice which has formed from the freezing either of water droplets intercepted by the pellets or of water resulting from the partial melting of the pellets.

Usually small hail is not easily crushable; when it falls on hard ground it bounces with an audible sound on impact. It always occurs in the form of showers.

4.2.1.9. *Diamond dust:*   precipitation which falls from a clear sky as very small ice crystals, often so tiny that they appear to be suspended in the air. The crystals are visible mainly when they glitter in the sunshine, giving rise to generally well-marked halo phenomena. Diamond dust can be observed in polar regions and in the interior of continents in winter, especially in clear, calm and cold weather. It is not often observed in the United Kingdom.

4.2.1.10. *Fog:*   a suspension of very small, usually microscopic, water droplets in the air, reducing visibility at the earth's surface to less than 1000 m When sufficiently illuminated, individual fog droplets are frequently visible to the naked eye; they are often seen to be moving in a somewhat turbulent manner.

The conditions resulting from the simultaneous occurrence of fog and heavy air pollution in urban and industrialized areas are widely referred to as 'smog' (smoke and fog).

The air in fog usually feels raw, clammy or wet, and the associated relative humidity is generally near 100 per cent.

Ice fog: a suspension of numerous minute ice particles in the air, reducing the visibility at the earth's surface to less than 1000 m. It forms when water vapour (mainly resulting from human activities) is introduced into the atmosphere at temperatures below -30 °C. This vapour condenses, forming droplets which freeze rapidly into ice particles having no well-defined crystalline form. Owing to their lack of form, these particles do not produce a halo which is only produced in ice fog when it contains diamond dust.

Shallow fog: fog lying on the surface of the ground or the sea, the depth of fog being below eye-level (about 1·8 m on land or 10 m at sea), with a visibility of 1000 m or more above the fog layer.

4.2.1.11. *Mist:*   a suspension in the air of microscopic water droplets or wet hygroscopic particles, reducing the visibility at the earth's surface. The term 'mist' is used in weather reports when the associated visibility is 1000 m or more and the corresponding relative humidity is 95 per cent or more but is generally lower than 100 per cent. Mist forms a generally fairly thin greyish veil which covers the landscape.

4.2.1.12. *Drifting snow and blowing snow:*   an aggregate of snow particles raised from the ground by a sufficiently strong and turbulent wind. The occurrence of these hydrometeors depends on the wind conditions (speed and gustiness) and the state and age of the surface snow.

(*a*) *Drifting snow:* an aggregate of snow particles raised by the wind to small heights above the ground and which veils or hides small obstacles. The visibility is not sensibly diminished at eye-level. (Eye-level is defined as 1·8 m above the ground.) The motion of the snow particles is more or less parallel to the ground.

(*b*) *Blowing snow:*   an aggregate of snow particles raised by the wind to moderate or great heights above the ground. The concentration of the snow particles may sometimes be sufficient to veil the sky and even the sun. Vertical visibility is diminished according to the intensity of the phenomenon; horizontal visibility at eye-level is generally very poor. When the phenomenon is severe it is sometimes difficult to distinguish whether snow is falling at the same time. Generally the blowing snow will be smaller in size than falling snow and at night, in particular, this difference may be detected by watching the snow passing through a light beam from either a torch or a cloud searchlight.

4.2.1.13. *Spray:*   an aggregate of water droplets torn by the wind from the surface of an extensive body of water, generally from the crests of waves, and carried up a short distance into the air. When the water surface is rough, the droplets may be accompanied by foam.

4.2.1.14. *Deposit of fog droplets:*   deposit of non-supercooled fog (or cloud) droplets on objects, the surface temperature of which is above 0 °C.

This hydrometeor is observed especially in high areas where orographic clouds are frequent. When the phenomenon is marked, the droplets run together and drip on to the ground.

4.2.1.15. *Dew:*   a deposit on objects of water drops produced by the direct condensation of water vapour from the surrounding clear air. It can form in two ways:

(*a*) *Dew proper* is formed when exposed surfaces are sufficiently cooled, generally by nocturnal radiation, to bring about the direct condensation of the water vapour from the surrounding air. Dew is deposited ordinarily on objects at or near the ground, mainly on their horizontal surfaces. It is observed especially during the warmer part of the year when the air is calm and the sky is clear.

(*b*) *Advection dew* is formed when exposed surfaces are sufficiently cold to bring about direct condensation of the water vapour contained in the air coming into contact with those surfaces, usually through a process of advection. Advection dew is deposited mainly on vertical exposed surfaces and is observed especially during the colder part of the year when relatively warm damp air suddenly invades a region after a period of moderate frosts.

The term 'white dew' is used for a deposit of white frozen dewdrops.

Dew should not be confused with the deposit of droplets from low fog on exposed surfaces, or, in the case of plants, with the droplets they may exude by a process known as guttation. This exudation of liquid water from the leaves of plants under warm, moist soil conditions often takes place at the same time as the deposit of dew, but can occur quite separately.

4.2.1.16. *Hoar frost:*   a deposit of ice which forms on objects and is generally crystalline in appearance, and produced by the direct sublimation of water vapour from the surrounding air. There are two types:

(*a*) *Hoar frost proper* is a deposit of ice which generally assumes the form of scales, needles, feathers or fans. It forms on objects whose surfaces have been sufficiently cooled, generally by nocturnal radiation, to bring about the direct sublimation of the water vapour contained in the air.

(*b*) *Advection hoar frost* is a deposit of ice, generally in crystalline form, which forms on objects whose surfaces are sufficiently cold to bring about the direct sublimation of the water vapour contained in the air coming into contact with these surfaces through a process of advection. It is deposited mainly on exposed vertical surfaces, usually when relatively warm damp air invades a region after a long period of hard frosts.

4.2.1.17. *Rime:*   a deposit of ice generally formed by the freezing of supercooled fog or cloud droplets on objects whose surface temperature is below or slightly above 0 °C. The thickness of the layer of rime should be measured and noted when practicable.

There are three species of rime: soft rime, hard rime and clear ice:

(*a*) *Soft rime* (see Plate XVI) is a fragile deposit consisting mainly of thin needles or scales of ice. It mainly forms when the ambient air temperature is lower than −8 °C. At and near the ground it is deposited under calm or light wind conditions on all sides of exposed objects. The deposit can easily be dislodged by a slight shake.

(*b*) *Hard rime* is a granular deposit, usually white, adorned with crystalline branches of grains of ice which are more or less separated by entrapped air. Hard rime mainly forms at air temperatures between −2 and −10 °C when supercooled water droplets rapidly freeze more or less individually and leave gaps between the frozen particles.

At and near the ground it is deposited mainly on the surface of objects exposed to a moderate or strong wind. In the windward direction the deposit may build up to form a thick layer. The deposit is rather adhesive but can be scratched off objects.

(*c*) *Clear ice* is a smooth, compact deposit which is usually transparent. It is fairly amorphous with a ragged surface and structurally resembles glaze. In nearly every case the temperature of the ambient air is between 0 and −3 °C, and clear ice is formed by the slow freezing of the supercooled water droplets. Before freezing, these water droplets penetrate the gaps between other fragments of ice.

At and near the ground, clear ice is deposited mainly on the surface of objects exposed to the wind; it is most likely to occur in mountain regions. The deposit is very adhesive and can only be removed from objects by being broken or melted.

4.2.1.18. *Glaze:* a smooth, compact deposit of ice, generally transparent, formed by the freezing of supercooled drizzle droplets or raindrops on objects the surface temperature of which is below or slightly above 0 °C. It covers all parts of surfaces exposed to precipitation, and is generally fairly homogeneous and resembles clear ice.

At and near the ground it is observed when drizzle droplets or raindrops fall through a layer of air below 0 °C of sufficient depth. It forms by the slow freezing of the water remaining in the liquid state after the cessation of supercooling, and the water is therefore able to penetrate the crevices between the particles of ice before freezing (see Plate XVI).

The deposit of ice formed by the freezing of fog not supercooled at the time of impact with objects the temperature of which is well below 0 °C, is also known as glaze.

Note: Glaze on the ground must not be confused with ground ice, which is formed when

(*a*) water from a precipitation of non-supercooled drizzle droplets or raindrops *later* freezes on the ground, or

(*b*) snow on the ground freezes again after having completely or partly melted, or

(*c*) snow on the ground is made compact and hard by traffic.

4.2.1.19. *Spout:* it consists of an often violent whirlwind, revealed by the presence of a cloud column or inverted cloud cone (funnel cloud) protruding from the base of a cumulonimbus, and of a 'bush' composed of water droplets raised from the surface of the sea or of dust, sand or litter raised from the ground.

The axis of the funnel cloud can be vertical, inclined, or sometimes sinuous. Not uncommonly, the funnel merges with the 'bush'.

The air in the whirlwind rotates rapidly, most often in a cyclonic sense: a rapid rotary movement may also be observed outside the funnel and the 'bush'. Further away the air is often calm.

The diameter of the cloud column, which is normally of the order of 10 metres, may in certain regions occasionally reach some hundreds of metres. Several spouts may sometimes be observed connected with a single cloud.

Spouts are often very destructive in North America (where they are called tornadoes). They may leave a path of destruction up to 5 km wide and several hundred kilometres long.

Weak spouts are occasionally observed under cumulus clouds.

When a spout is observed, the height, diameter, sense of rotation and path of the cloud funnel (also called 'tuba') should be noted as far as possible. It is also useful to obtain information about any damage done.

4.2.2. Lithometeors. A lithometeor is a meteor consisting of an aggregate of particles, most of which are solid or non-aqueous. The particles are (*a*) more or less suspended in the air (haze, dust haze and smoke) and consist of very small dust particles, or sea-salt particles or combustion products, or (b) lifted from the ground by the action of the wind (drifting or blowing dust or sand, dust whirl or sand whirl, duststorm or sandstorm).

The most common lithometeors are described below.

4.2.2.1. *Haze:* a suspension in the air of extremely small, dry particles invisible to the naked eye and sufficiently numerous to give the air an opalescent appearance.

There is no upper or lower limit to the horizontal visibility in the presence of which haze may be reported. Haze imparts a yellowish or reddish tinge to distant bright objects or lights seen through it, while dark objects appear bluish, mainly as a result of scattering of light by the haze particles. These particles may have a colour of their own which also contributes to the coloration of the landscape.

4.2.2.2. *Dust haze:* a suspension in the air of dust or small sand particles raised from the ground prior to the time of observation by a duststorm or sandstorm. The duststorm or sandstorm may have occurred either at or near the station or far from it. Dust haze is reported only when there is an absence of turbulent or strong wind sufficient to raise dust from the ground at the time of observation.

4.2.2.3. *Smoke:* a suspension in the air of small particles produced by combustion.

This lithometeor may be present either near the earth's surface or in the free atmosphere. It is often the visible product of the incomplete combustion of coal; in the United Kingdom the frequency of this phenomenon has been greatly reduced by the introduction of 'smokeless zones'. Coal smoke consists of mainly carbon and hydrocarbon particles of very small size (about $0.1 \mu m$) which remain in the air, on average, for one or two days. Viewed through smoke, the sun appears very red at sunrise and sunset, and shows an orange tinge when high in the sky. Smoke in extensive layers originating from fairly near forest fires scatters the sunlight and gives the sky a greenish-yellow hue. Evenly distributed smoke from very distant sources generally has a light greyish hue.

When visibility is impaired and the cause can be definitely attributed to smoke, then 'visibility reduced by smoke' is reported (as is provided for in the present-weather specification in the *Handbook of weather messages,* Part II, *Abbreviated weather reports* and in Metform 3100A).

4.2.2.4. *Drifting and blowing dust or sand:* an aggregate of particles of dust or sand raised, at or near the station, from the ground to small or moderate heights by a sufficiently strong and turbulent wind.

The wind conditions (speed and gustiness) necessary to produce these lithometeors depend on the nature and state of the ground, for example on the degree of dryness of the ground.

(a) *Drifting dust or drifting sand:* dust or sand raised by the wind to small heights above the ground. The visibility is not sensibly diminished at eye-level (1.8 m above the ground). The motion of the particles of dust or sand is more or less parallel to the ground.

(b) *Blowing dust or blowing sand:* dust or sand raised by the wind to moderate heights above the ground. The horizontal visibility at eye-level is sensibly reduced. The concentration of the particles of dust or sand may sometimes be sufficient to veil the sky and even the sun.

4.2.2.5. *Duststorm or sandstorm:* an aggregate of particles of dust or sand energetically lifted to great heights by a strong and turbulent wind.

Surface visibility is reduced to low limits; the qualification for inclusion in a British report is visibility below 1000 m.

Duststorms or sandstorms generally occur in areas where the ground is covered with loose dust or sand; sometimes, after having travelled over great distances, they may be observed over areas where neither dust nor sand covers the ground.

The forward portion of a duststorm or sandstorm may have the appearance of a wide and high wall which advances more or less rapidly. Walls of dust or sand often accompany a cumulonimbus which may be hidden by the dust or sand particles. These walls may also occur without any clouds along the forward edge of an advancing cold air mass.

In a slight or moderate duststorm or sandstorm the visibility is less than 1000 m but not below 200 m and the sky is not usually obscured. In a severe duststorm or sandstorm the visibility is reduced below 200 m and the sky is usually obscured.

4.2.2.6. *Dust whirl or sand whirl (dust devil):* an aggregate of particles of dust or sand, sometimes accompanied by small litter, raised from the ground in the form of a whirling column of varying height with a small diameter and an approximately vertical axis.

These lithometeors occur when the air near the ground is very unstable, as for instance when the dust or sand surface is strongly heated by the sun.

4.2.3. Electrometeors. An electrometeor is a visible or audible manifestation of atmospheric electricity. Electrometeors either correspond to discontinuous electrical discharges (thunder, lightning) or occur as more or less continuous phenomena (St Elmo's fire, polar aurora, both described in Chapter 11).

4.2.3.1. *Thunderstorm:* one or more sudden electrical discharges, manifested by a flash of light (lightning) and a sharp or rumbling sound (thunder).

Thunderstorms are associated with convective clouds and are most often, but not necessarily, accompanied by precipitation at the ground.

In recording observations, a thunderstorm is regarded as being at the station from the time thunder is first heard, whether or not lightning is seen or precipitation is occurring at the station. A thunderstorm is reported in present weather if thunder is heard within the normal observational period (10 minutes) preceding the time of the report. The World Meteorological Organization convention is that a thunderstorm is regarded as having ceased at the time of the last audible thunder and the cessation is confirmed if thunder is not heard for 10–15 minutes after this time. However, practice for Meteorological Office stations is that an observer should not report 'thunderstorm at the time of observation' when the cumulonimbus has already passed the station even though it be less than 10 minutes since he heard the last peal of thunder. When he is not sure that the storm has passed he should report a 'thunderstorm at the time of observation' if thunder has been heard within the last 10 minutes.

4.2.3.2. *Lightning*: a luminous manifestation accompanying a sudden electrical discharge which takes place from or inside a cloud or, less often, from high structures on the ground or from mountains.

Three main types of lightning can be distinguished:

(a) *Ground discharges* (popularly called thunderbolts or forked lightning). This type of lightning occurs between cloud and ground; it comprises a leader stroke which follows a tortuous downward course and is often branched, from a distinct main channel (streak or ribbon lightning). One of these branches establishes an ionized path to the ground up which a return stroke then passes. It is the return stroke which carries the main energy transfer. Additional leader and return strokes may follow. A luminous globe has occasionally been observed, soon after a discharge to ground. This globe, the dimension of which has been reported to be generally between 10 and 20 cm, but is said sometimes to reach 1 m, is known as ball lightning. It moves slowly in the air or on the ground and usually disappears with a violent explosion.

(b) *Cloud discharges* (popularly called sheet lightning). This type of lightning takes place within the cumulonimbus; it gives a diffuse illumination without a distinct channel being usually seen. This type of lightning includes the so-called heat lightning, consisting of diffuse light flashed from distant thunderstorms seen near the horizon.

(c) *Air discharges*. This type of lightning occurs in the form of sinuous discharges, often ramified but with a distinct main channel, passing from a cumulonimbus to the air and not striking the ground. It frequently includes a long quasi-horizontal part. The name 'streak lightning' is also applied to this type of lightning.

4.2.3.3. *Thunder*: a sharp or rumbling sound which accompanies lightning, caused by a sudden heating and expansion of the air along the path of the lightning. The distance of a lightning flash may be roughly estimated from the interval between seeing the flash and hearing the thunder, counting 1 km for every three seconds. The long duration of thunder compared with the associated lightning flash is explained by the different distances travelled by the sound from different parts of the flash and by echoing from mountain sides. Echoing causes intensity variations; however, variations also arise from the multiple and tortuous nature of many lightning strokes.

Thunder is seldom heard at distances greater than 20 km. Owing to refraction of sound waves in the lower atmosphere, thunder is sometimes inaudible at distances much less than 20 km, especially when the initiating lightning flash is not to ground.

4.2.4. Other phenomena. Photometeors are dealt with in Chapter 11, together with some additional electrometeors which may be more accurately considered as atmospheric rather than weather phenomena. However, there are some phenomena not listed in the *International cloud atlas* as meteors which should be recorded in weather observations and these are described below.

4.2.4.1. *Squall*: a strong wind that rises suddenly, lasts for at least a minute and then dies away comparatively suddenly (see 5.1.2.2, page 82). Squalls are frequently associated with cumulonimbus or with the passage of cold fronts. In the latter circumstances they occur in a line (the line of the front) and are typically accompanied by a sharp fall in temperature, a veer of wind, a rise of relative humidity, and a roll-shaped cloud with horizontal axis. These phenomena are known collectively as a line-squall.

4.2.4.2. *Snow lying*: snow lying means, in general, snow covering the ground, either completely or in patches. For the purpose of climatological returns a day with snow lying is one in which snow covers one half or more of the ground of an open area representative of the station at the morning climatological hour of observation (0900 GMT in the United Kingdom). The ground of an open area representative of the station should be taken to include the open, fairly flat ground easily visible from the station and not differing from it in altitude by more than 30 m. In judging whether half the ground is covered, no account should be taken of bare patches under trees or areas occupied by rivers, ponds, etc. (see 6.1.2, page 95).

There are provisions for reporting the character and depth of snow cover in the codes given in the *Handbook of weather messages*, Part II.

4.3. PRESENT WEATHER

The observation of present weather necessitates noting the state of the sky, or the change of the state of sky, and the phenomena occurring at the station, or within sight of the station, at the time of observation. Relevant coding procedures and instructions for climatological stations are given in Metform 3100A (Supplement to the Pocket Register), and for synoptic stations in the *Handbook of weather messages*, Parts II and III, and *Abbreviated weather reports*. These procedures provide for the occasions when precipitation or fog, or a thunderstorm, has ceased during the past hour.

4.3.1. Definition of terms.

4.3.1.1. *Precipitation* is the general term given to water drops or ice particles formed at a higher level and falling to or towards the ground. It includes rain, drizzle, sleet, snow, snow pellets, snow grains, ice pellets, hail, small hail and diamond dust. The term does not include drifting or blowing snow, duststorms or sandstorms, or dust whirls and sand swirls, which are not formed at a higher level but merely raised by the wind. Surface condensation phenomena such as dew, hoar frost and rime which may contribute to the catch of a rain-gauge are not classed as precipitation under the heading of present weather.

4.3.1.2. *At the station* means literally at the place where the observations are normally taken. For thunderstorms, however, it is not necessary for the disturbance to be immediately overhead. The storm is regarded as occurring at the station if thunder is heard (see 4.2.3.1).

4.3.1.3. *Within sight* or *at a distance* means that although the weather in question is not actually occurring at the station, it is seen to be occurring in the vicinity. When applied to precipitation, the phrase should refer only to cases where the precipitation is seen to reach the ground, or to evaporate at a low level; it should not be used to describe virga (see 2.3.4) associated with cirrocumulus and altocumulus floccus.

4.3.1.4. *During the past hour* or *in the past hour* refers to the approximate one-hour period between the actual time of an observation of an element and the previous actual time of observation. Even if hourly observations are not made it still refers to the one-hour period before the actual time of observation.

4.3.1.5. *At the time of observation* is to be interpreted as the actual time of observation of the element concerned.

4.3.2. Continuity of precipitation.   In observing and reporting present weather, precipitation is specified as intermittent, continuous or as showers. Precipitation, other than showers, is reported as continuous when it has continued for at least 60 minutes without a break. All other precipitation, except showers, is reported as intermittent. If precipitation stops before an observer has completed his observation, he will report 'precipitation in the past hour' in the present-weather code.

Continuous precipitation falls from layer clouds which usually form a dense covering to the whole sky.

Intermittent precipitation falls from layer clouds which almost always cover the whole sky, though there may be considerable variations in the thickness and opacity of the layer. Occasional brightening or change in coloration of the sky, without a definite break in the cloud, is not unusual and lifting of the cloud base may occur at times.

4.3.2.1. *Showers* are said to occur when solid or liquid precipitation falls from convection (cumuliform) clouds. The amount of cloud usually varies greatly during the course of an hour or so, and often within much briefer periods. Convection clouds are usually seen either building up at a station before the shower begins or approaching the station before the precipitation reaches the station. When a well-developed shower cloud is over a station it may for a time cover the whole sky, but after the shower there is usually a partial, and sometimes a complete, clearance of the sky; the cloud may be seen to decay or to move away after the precipitation at the station has ceased. It is rare for a true shower to take the form of drizzle, while hail is always a shower type of precipitation. Rainbows are often associated with showers.

4.3.3. Intensity of precipitation.   No international definitions have been agreed for the terms 'slight', 'moderate' and 'heavy' used in describing the intensity of precipitation. The Meteorological Office classifications of intensity of rain are given in 4.3.3.1. See 4.3.3.2. for showers (for which there is an additional classification, namely 'violent'), 4.3.3.3 for drizzle, and 4.3.3.4 for snow. Guidance on reporting the intensity of mixed precipitation is given in 4.3.3.5. The classifications refer to the rate of fall of the precipitation and not to the total amount.

If the intensity of any precipitation alters during the course of making an observation the observer should report the latest intensity.

4.3.3.1. *Intensity of rain.*

(*a*) Slight rain is rain of low intensity; it may consist of scattered large drops or more numerous smaller drops. The rate of accumulation on the ground is such that puddles form only slowly, if at all. The rate of accumulation in a rain-gauge is not more than 0·5 mm per hour.

(*b*) Moderate rain is rain falling fast enough to form puddles rapidly, to make downpipes flow freely, and to give some spray over hard surfaces. The rate of accumulation in a rain-gauge is between about 0·5 and 4 mm per hour.

(*c*) Heavy rain is a downpour which makes a roaring noise on roofs, forms a misty spray of fine droplets by splashing on road surfaces etc., and accumulates in a rain-gauge at a rate of more than 4 mm per hour.

At a station equipped with either a Meteorological Office tilting-siphon rain recorder, or tipping-bucket gauge with 750 cm² collector and an indicating counter, the observer can make use of either of these instruments in assessing the intensity of rainfall.

As noted in 4.3.3 above, an observer must report the intensity he last observed and this will not necessarily be the intensity shown either by the rain-recorder chart when it was examined, or by the final counter reading if a tipping-bucket gauge is used to assess intensity, as suggested in the previous paragraph. By noting these recorded intensities, and noting any changes which occur while making his observation, it should be possible for him to assess the intensity last observed. On occasions when the assessment is on the borderline of two intensities the higher intensity should be reported.

As a guide, Figure 5 shows examples of the slope of the trace on the chart (Metform 4423A) of a Meteorological Office tilting-siphon rain recorder in prolonged steady rain of slight, moderate and heavy intensity (at A, B and C respectively).

At a station equipped with a tipping-bucket gauge with 750 cm² collector and an indicating counter, the observer notes the counter reading at the beginning of the observation, and a further reading taken precisely six minutes later. A difference in the readings of 3 or more indicates that the rainfall

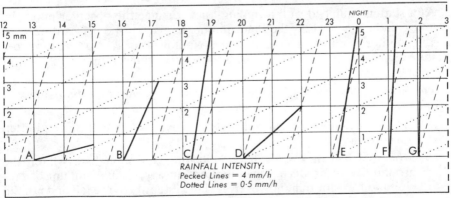

Figure 5.   Sample traces on a tilting-siphon rain-recorder chart

has been heavy, averaged over the six minutes. A difference of 2 or less is inconclusive, so the observer will have to use his own judgement as to whether the intensity of the rain is slight or moderate.

4.3.3.2. *Intensity of showers.* The scales of intensity quoted above for slight, moderate and heavy rain need to be increased when applied to rain showers because the convective processes which cause showers are generally more vigorous than the frontal or orographic uplift of air associated with continuous precipitation. The following descriptions and rate of fall should be taken as a guide:

(*a*) Slight showers. These vary from scattered drops to drops falling fast enough to form puddles. The rate of accumulation in a rain-gauge is less than about 2 mm per hour.

(*b*) Moderate or heavy showers. These form puddles rapidly and in the heavier precipitation make a roaring noise on roofs and produce a misty spray when the drops strike a hard surface. Visibility is often impaired significantly. The rate of accumulation in a rain-gauge of a moderate shower is about 2–10 mm per hour; in a heavy shower it is about 10–50 mm per hour.

(*c*) A violent shower is one in which the intensity is exceptional for the British Isles, although not uncommon in the tropics. The rate of accumulation in a rain-gauge is greater than about 50 mm per hour.

In Figure 5 examples of the slope of the trace on a chart of a Meteorological Office tilting-siphon rain recorder in showers of slight, moderate, heavy and violent intensity are shown at D, E, F and G respectively. In the same way as described in 4.3.3.1, the intensity as shown by the recorder may be used as a guide but with a degree of caution, for three reasons. Firstly, the slope of the trace is often steep and is therefore difficult to judge correctly. Secondly, a uniform intensity is often not maintained long enough for the trace to appear on the chart as a straight line. Thirdly, and this applies generally, conditions are not favourable for precise observation while the chart is still on the drum of the instrument.

The window of the recorder should not be opened if the precipitation is heavy because water splashing from the ground may ruin the chart. If the chart gets wet its adjustment on the drum is likely to be affected and the subsequent part of the trace may be falsified.

At a station equipped with a tipping-bucket gauge the observer notes the counter reading in precisely the same manner as given in 4.3.3.1. A difference in the counter readings, averaged over six minutes, of 2, 3 or 4 counts per minute indicates a moderate shower, 6–24 a heavy shower, and more than 24 a violent shower.

The notes below should be used as a guide for reporting hail showers. (Guidance for reporting snow showers is given in 4.3.3.4.)

(*a*) Slight hail consists of sparse hailstones, usually of small size and often mixed with rain.

(*b*) Moderate hail means a fall of hail abundant enough to whiten the ground and to produce, when melted, an appreciable amount of precipitation. As for rain, however, it is the intensity of the fall and not the total amount which determines the classification.

(*c*) Heavy hail is exceptional in the British Isles and includes at least a proportion of stones exceeding 6 mm in diameter. Ground crops are damaged and leaves knocked off trees. Glass may be broken in greenhouses, garden frames, etc.

4.3.3.3. *Intensity of drizzle.* As defined in 4.2.1.2, drizzle is fairly uniform precipitation composed exclusively of fine drops of water very close to one another. The diameter of the drops is usually less than 0·5 mm.

The effect of their individual impact on water surfaces is imperceptible. Continuous drizzle may produce a run-off from roofs and road surfaces, and a rate of accumulation in the rain-gauge equal to or exceeding that produced by slight rain. Moderate or heavy drizzle will have a marked effect on the visibility.

The estimation of intensity of drizzle is especially difficult. However, the classification given below should be used as a guide.

(*a*) Slight drizzle can readily be detected on the face and, for example, on the windscreen of a car, but produces very little run-off from road surfaces or roofs.

(*b*) Moderate drizzle causes windows and road surfaces to stream with moisture.

(*c*) Heavy (dense) drizzle impairs visibility significantly, and accumulates in the rain-gauge at a rate up to 1 mm per hour.

Drizzle frequently occurs in association with mist or fog, but may produce poor visibility in otherwise clear air.

4.3.3.4. *Intensity of snow and snow showers.* The classification of intensity of snowfall is qualitatively made as indicated below:

(*a*) Slight when the flakes are sparse and usually small. In calm weather the rate of accumulation of the snow cover does not exceed 0·5 cm per hour.

(*b*) Moderate when the snowfall consists of usually large flakes falling sufficiently thickly to impair visibility substantially. The snow cover increases in depth at a rate up to 4 cm per hour.

(*c*) Heavy when visibility is reduced by the falling snow to a low value and the snow cover increases at a rate exceeding 4 cm per hour.

In the above statements as to the depth of snow, it is assumed that the temperature is below freezing point so that melting does not occur, and that drifting is not taken into account.

It should be noted that the term 'blizzard' is not recognized as a description of the weather. However, it is a term used during the occurrence of severe winter conditions and the following definitions have been accepted for use within the United Kingdom in order that some uniformity of practice is established in the use of the term.

Blizzard: the simultaneous occurrence of moderate or heavy snowfall and winds of at least force 7 (28 knots) which causes drifting snow and a reduction of the visibility to 200 m or less.

Severe blizzard: the simultaneous occurrence of moderate or heavy snowfall and winds of at least force 9 (41 knots) which causes drifting snow and a reduction of the visibility to near zero.

These terms will not be applied to passing snow showers but only if a wide area is affected and the conditions last long enough to cause serious interference to human mobility, or the disruption of communications.

The hydrometeors snow grains (see 4.2.1.5), ice pellets (see 4.2.1.6) and diamond dust (see 4.2.1.9) are all reported without classification of intensity. However, any relevant observations of intensity should be noted in the remarks column of the observation Register.

4.3.3.5. *Intensity of mixed precipitation.*   When mixed precipitation occurs such as drizzle and rain, or hail and snow, the intensity of each type is not given separately. The intensity of the heaviest precipitation is used to denote the intensity of the mixture. If any doubt exists on this point, the best estimate of the following, in order of precedence shown, should be reported: (1) hail, (2) snow, (3) rain, (4) drizzle.

4.4. PAST WEATHER

Details of the procedures and use of the figure code for reporting past weather are given in the *Handbook of weather messages*, Parts II and III and *Abbreviated weather reports*.

4.4.1. Climatological stations.   In general, climatological stations are not asked to report past weather as such, although a weather diary is required (see 4.1.1, page 57); it should however be noted that the Health Resort stations do report past weather in their evening coded messages (see 1.5.2, page 8).

4.4.2. Synoptic stations.   At synoptic stations the interval covered by the past-weather description is six hours in reports at 0000, 0600, 1200 and 1800 GMT, three hours in reports at 0300, 0900, 1500 and 2100 GMT, and one hour in reports at other times. Particular exceptions to this rule are detailed in the *Handbook of weather messages*, Part III.

4.5. BEAUFORT LETTERS AND INTERNATIONAL SYMBOLS

Beaufort letters are used by the Meteorological Office to provide a continuous record of the weather in a brief form. The code of letters indicating the state of the weather, past or present, was originally introduced by Admiral Sir Francis Beaufort early in the nineteenth century for use at sea, but they are equally convenient for use on land. Since his day the code has been substantially modified.

Beaufort letters can also be used as a means of describing the weather over a period of time for transmission either by telephone or teleprinter. This method is used by selected stations and by stations which participate in the Health Resort Scheme (see page 79). The *Handbook of weather messages*, Part III, gives instructions on their use in the Daily Register.

The appropriate letters from the first column of the table given in 4.5.1 are selected for entry in the relevant column of the Register. The second column gives the international symbols and some others. These provide a convenient way of noting at any time, in the limited space of the remarks column of the

Register, the occurrence of any phenomena which might otherwise go unrecorded. Such small additions provide some background information to the record of a station.

4.5.1. Table of Beaufort letters, international symbols for meteors, and some additional symbols.

State of sky

b		Total cloud amount 0-2/8
bc		Total cloud amount 3/8-5/8
c		Total cloud amount 6/8-8/8
o		Uniform thick layer of cloud completely covering the sky: 8/8

Hydrometeors

r	•	rain
r	∾	freezing rain
d	,	drizzle
d	∾	freezing drizzle
s	✳	snow
h	✻	snow pellets
h	↔	diamond dust
h	▲	hail
h	△	small hail
h	⬠	ice pellets
sh	⟋	snow grains
f	≡	fog
f	⇌	ice fog
fe	≡:	*wet fog
fg/fs	= =	*patches of shallow fog over land/sea
fg/fs	=⸺=	*more or less continuous shallow fog over land/sea
m	=	mist
ks	⊣⊦	drifting and blowing snow
ks	⊣⊦	drifting snow
ks	⊣⊦	blowing snow

	(spray)	spray
w	⌐	dew
w	⌐	advection dew
w	━	white dew
x	∟	hoar frost
x	⌐	advection hoar frost
	∨	rime
	∀	soft rime
	∀	hard rime
	⩜	clear ice
	∿	glaze
	)(	spout

(mixed precipitation)

dr	;	drizzle and rain
rs	⁎	rain and snow (sleet)
hs	▲⁎	hail and snow
hr	▲	hail and rain

Lithometeors

z	∞	haze
	S	dust haze
	∿	smoke
	$	drifting and blowing dust or sand
	$	drifting dust or sand
	$	blowing dust or sand
kz	⭲	duststorm or sandstorm
	⭲	wall of dust or sand
	⑀	dust whirl or sand whirl (dust devil)

Photometeors

	⊕	solar halo
	⌓	lunar halo

◐		solar corona
∪		lunar corona
⊘		irisation
⊗		glory
⌒		rainbow
⌒		fog-bow
◎		Bishop's ring
⋈		mirage
◭		*zodiacal light

Electrometeors

tl	⃔	thunderstorm
l	⃕	lightning
	⃕	St Elmo's fire
	⌒	polar aurora

Miscellaneous

j		phenomenon within sight of but not at the station
e		wet air, without rain falling
y		dry air (less than 60 per cent relative humidity)
u		ugly threatening sky
v		abnormally good visibility
p	▽	shower (used in combination with the type of precipitation)

Surface wind

g	⧹⧹⧹	*gale, mean speed 34-47 knots over a period of 10 minutes or more
G	∟	*storm, mean speed 48 knots or more over a period of 10 minutes or more
q	⅄	*squall
kq	⅄	*line squall

---

*Not internationally accepted symbols.

4.5.2. Recording Beaufort letters. When Beaufort letters are used for recording the weather in the Register, the phenomena are recorded in the order in which they occur. Beaufort letters for the weather during the full course of an observation will be allocated to the period preceding the observation. If the weather at the time of the observation continues into the next observational period the appropriate Beaufort letters should be included in both observations.

In a sequence of Beaufort letters a change from one set of conditions to another is indicated by the insertion of a comma between the sets.

When several phenomena occur simultaneously they are recorded in the following order:

(1) state of sky   (4) atmospheric obscurity

(2) thunderstorm   (5) other phenomena.

(3) precipitation

4.5.2.1. *State of sky.*   The state of sky is required to be recorded in every combination of Beaufort letters except when the sky is obscured, when no letter is required. Note that the state of sky refers to the amount of cloud cover. The use of the Beaufort letter 'u', an ugly threatening sky, is not used in this context.

4.5.2.2. *Thunderstorm.*   A thunderstorm is regarded as being at the station from the time thunder is first heard whether or not lightning is seen or precipitation is occurring at the station.

4.5.2.3. *Precipitation*:   type, intensity, continuity. In recording precipitation in Beaufort letters account must be taken of the variations that can arise in type, intensity, continuity, or combinations of some or all of them. These are dealt with in detail below.

(*a*) *Type.*   The type of precipitation is indicated by the appropriate letter, or combination of letters if there is a mixture of precipitation. For example:

$$d = \text{drizzle}; \quad r = \text{rain}; \quad dr = \text{drizzle and rain}.$$

If the precipitation is of the showery type (falling from convective cloud), the prefix 'p' is used in combination with the type of precipitation. For example:

$$pr = \text{shower of rain}; \quad ps = \text{shower of snow}.$$

(*b*) *Intensity.*   The intensity of precipitation is recorded in four categories: slight, moderate, heavy and violent. These are indicated in the following manner:

(1) Slight: by the addition of the subscript 'o' to a small Beaufort letter. For example:

$$r_o = \text{slight rain}; \quad s_o = \text{slight snow}; \quad pr_o = \text{slight shower of rain}.$$

(2) Moderate: by a small Beaufort letter. For example:

$$r = \text{moderate rain}; \quad s = \text{moderate snow}; \quad pr = \text{moderate shower of rain}.$$

(3) Heavy: by a capital Beaufort letter. For example:

$$R = \text{heavy rain}; \quad S = \text{heavy snow}; \quad pR = \text{heavy shower of rain}.$$

(4) Violent: by the addition of the subscript '2' to the capital Beaufort letter. For example:

$$pR_2 = \text{violent shower of rain.}$$

When mixed precipitation occurs, such as drizzle and rain, or rain and snow, the intensity of each type is not given separately, but the intensity of the heaviest precipitation is used to denote the intensity of all the other types in the mixture. For example:

slight drizzle and moderate rain = dr.

The intensity of a thunderstorm is judged by the intensity of the thunder and lightning, whilst the intensity of the precipitation in the storm is indicated separately. For example:

TLr_o = heavy thunderstorm with slight rain
tl_oR = slight thunderstorm with heavy rain.

When showers are reported, the qualification of intensity is given to the precipitation, but not to the shower prefix 'p'.

(c) *Continuity.* Precipitation falling from layer cloud is described by letters referring to the continuity as well as to the type of intensity in accordance with the following rules:

(1) Intermittent precipitation: the Beaufort letters indicating the type and intensity of the precipitation are prefixed by the letter 'i'. For example:

ir_o = intermittent slight rain
iS = intermittent heavy snow
idr = intermittent moderate drizzle and rain.

The prefix indicates that there has been a break or breaks occurring at intervals of less than one hour in the overall period of the precipitation. Note that an individual break lasting one hour or more requires subsequent precipitation to be recorded as the commencement of another period.

(2) Continuous precipitation: the Beaufort letter(s) indicating the type and intensity of the precipitation are repeated. For example:

r_or_o = continuous slight rain
SS = continuous heavy snow
$d_or_od_or_o$ = continuous slight drizzle and rain.

The repetition indicates that the period of precipitation has lasted for at least one hour without a break.

(3) Precipitation not specified as intermittent or continuous: the Beaufort letter(s) indicating the type and intensity of the precipitation are used alone. For example:

R = heavy rain
d_o = slight drizzle
dr = moderate drizzle and rain.

This indicates that the period of precipitation has not lasted for one hour to qualify it as continuous, and that there have been no breaks to qualify it as intermittent.

(d) Changes of type and/or intensity. During a period of precipitation a change of type and/or intensity is indicated by successive use of letters descriptive of each new type or intensity. Repetition of letters to indicate continuity will be appropriate only when precipitation of one particular type and intensity has continued for at least one hour without a break. A change in type or intensity of continuous precipitation where the new type or intensity does not last for one hour will require the use of a single letter as described in (c)(3) above. At each change of type and/or intensity it is necessary to record all the appropriate letters in the order specified in 4.5.2 (page 76), and a comma is placed between each group of letters. For example:

$$cr_or_o, \; cr, \; cr_o, \; cd_or_o.$$

4.5.2.4. *Atmospheric obscurity*.   The appropriate Beaufort letter describing the cause of the obscuration is selected from the table in 4.5.1.

(*a*) Fog. Whenever the visibility is reduced to less than 1000 m and the obscuration is caused by fog (as defined in 4.2.1.10, page 60), the letter 'f' (fog) will be used down to and including 200 m, and the capital letter 'F' (thick fog) when the visibility is less than 200 m.

When patches of fog exist, i.e. the visibility varies with direction and the minimum visibility is less than 1000 m while the maximum is outside fog limits, the letters 'if' (intermittent fog) will be used.

The recording of continuity of fog is governed by the same rules as those governing the recording of continuity of precipitation in 4.5.2.3(c). Note also that continuity of precipitation and fog are assessed separately.

Changes of observed visibility in a period of fog which pass through the limit of 200 m will require the successive use of the letters 'f' and 'F', as appropriate. Repetition of either of these letters to indicate continuity will require one hour or more of fog in the particular range of density without a break; for example:

bcif, cf, ff, F.

(*b*) Mist (as defined in 4.2.1.11, visibility 1000 m or more) and

(*c*) Haze (as defined in 4.2.2.1).

In order to differentiate between obscuration caused by water droplets and dry particles, it is United Kingdom practice to use the following criteria: 'The obscuration is caused predominantly by water droplets if relative humidity is 95 per cent or more, and by dry solid particles if the relative humidity is less than 95 per cent. It is necessary, however, to preserve continuity in observations of present weather and it may, for this reason, be necessary not to apply the 95 per cent criterion too strictly on some occasions.'

4.5.2.5. *Description of phenomena within sight*.   The letter 'j' is used in combination with various other Beaufort letters to record phenomena occurring within sight of but not at the station; for example:

jp = precipitation within sight
jf = fog within sight
jks = drifting snow within sight
jkz = sandstorm or duststorm within sight.

No qualification of intensity or indication of the type of precipitation is applied to adjacent precipitation, 'jp', even though this might be surmised.

Adjacent precipitation is not used to describe a shower which was previously reported at the station and is still visible on the horizon.

4.5.3. Beaufort letters sent with coded messages.

In addition to the Health Resort stations which receive separate instructions (1.5.2 refers), selected stations make reports for the Press at 6 p.m. clock time and include Beaufort letters to describe the past weather in the morning and the afternoon. These stations are limited to two sets of five letter-spaces each for their Beaufort letter reports (see *Handbook of weather messages*, Part III). Exceptionally, the Beaufort letters 'e' and 'w' will be reported, when applicable, to explain the presence of measurable amounts of precipitation not otherwise accounted for. The letter 'z' will not be used.

When Beaufort letters are sent by teleprinter only capital letters can be used and the following conventions for indicating intensity apply:

slight	— the Beaufort letter preceded by 'N'
moderate	— the Beaufort letter
heavy, severe or intense	— the Beaufort letter preceded by 'A'
violent	— the Beaufort letter preceded by 'AA'.

Continuity is reported by prefixing each letter by the letter designated for intensity, for example:

NRNR = continuous slight rain
ARAR = continuous heavy rain.

Intermittent is indicated by the prefix 'I', for example:

INS = intermittent slight snow.

When showers are reported, the qualification of intensity is applied only to the letter indicating the precipitation, for example:

PAR = shower of heavy rain.

CHAPTER 5

SURFACE WIND

5.1. GENERAL

An adequate description of the wind generally requires both speed and direction to be specified. The effect of turbulence near the earth's surface is to produce rapid irregular changes in both the speed and direction of the wind. These fluctuations occur independently, over short intervals of time, and are referred to as 'gustiness'. Wind direction, speed and gustiness are generally best determined instrumentally, but when such determination is not practicable the wind direction and speed are estimated.

At present the official unit of horizontal wind speed in the United Kingdom is the knot (kn).* All speeds quoted in this chapter are in knots but, for some of them, metric equivalents are also given; these are the values that would be used in the United Kingdom if metres/second (m/s) became the official unit of measurement. The figures are not necessarily precise equivalents because, when speeds are expressed in knots, whole numbers are used, whereas speeds in metres/second are often given to the nearest tenth. For example, using the expressions in 5.1.2 below, the lower limit of wind speed in a gale would be 33·4 knots or 17·2 m/s; the former figure is rounded up to 34 but the direct conversion of that (17·5 m/s) is not used. (See table on page 87.)

The wind direction is always specified as that direction from which it is blowing. It is expressed in tens of degrees measured clockwise from true north, or in terms of the points of the compass. Scales relating direction from true north in whole degrees and points of the compass are shown in Figure 6.

The surface wind speed and direction are usually measured at a standard height of 10 metres above the ground in an open situation. Unless otherwise stated, the winds discussed in this chapter will refer to these standard conditions. Corrections to wind speed at other than the standard height are dealt with in 5.2.1.

5.1.1. Reports of surface wind. At each hour of observation an observer will report the surface wind direction in tens of degrees from true north and the wind speed in knots.

Additionally, at synoptic stations, any marked change of direction or speed, times of onset and cessation of gales, extreme speed and gusts will be reported in accordance with the requirements and reporting procedures which are to be found in the *Handbook of weather messages*, Part III. If an observer estimates either the wind direction on a 16-point scale (N, NNE, NE, etc.) referred to true north, or the wind speed on the Beaufort scale, then these are converted to degrees and knots respectively. The conversions

---

*1 knot = 1 international nautical mile/hour = 1852 metres/hour = 0·514 metres/second. The symbol kn is being increasingly used in international publications to avoid confusion with the kilotonne (kt).

can be made either by using Figure 6 and the Beaufort scale (pages 86–87) or, for climatological observers, the appropriate tables in Metform 3100A (see 1.5.1, page 8).

Observers at climatological stations should, whenever possible, note the occurrence of gales with the times of beginning and ending, and an estimate of the maximum speed attained. These times and comments should be entered in the remarks column of the Pocket Register and in the Weather Diary of the climatological return.

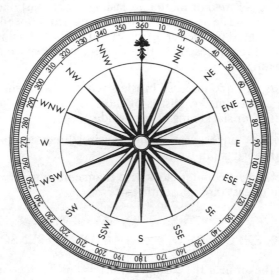

Figure 6. Specification of wind direction

5.1.2. Terminology. Specific terms are used to describe phenomena associated with the surface wind; they are:

5.1.2.1. *Gale.* A gale is defined as a surface wind of mean speed of 34 knots (17·2 m/s) or more, averaged over a period of 10 minutes. Terms such as 'strong gale', 'storm', etc. are also used to describe winds of 41 knots or greater (see pages 86–87).

Gale warnings are issued when mean speeds of 34 knots or more are expected but warnings are also issued if gusts of 43 knots (22 m/s) or more are expected with mean wind speeds of less than 34 knots, provided that these gusts are not isolated such as those accompanying squalls or thunderstorms.

Certain operational procedures have to be followed when gale-force winds are observed. For this purpose it is necessary to define the times of starting and ending of periods of gales.

A gale is said to be ended when there occurs a period of at least 10 minutes during which neither of the criteria used for gale warnings has occurred.

The time of commencement of a gale is the middle of the first 10-minute period in which the mean surface wind exceeded 34 knots or the time of the first gust exceeding 43 knots, whichever is the earlier. The time of cessation of a gale is the middle of the last 10-minute period in which the mean wind exceeded 34 knots or the time of the last gust exceeding 43 knots, whichever is the later.

5.1.2.2. *Squall.*   A squall is a strong wind that rises suddenly, lasts for at least a minute and then dies away comparatively suddenly. It is distinguished from a gust by its longer duration.

The following definition of a squall was adopted in April 1962 by the Third Session of the WMO Commission for Synoptic Meteorology:

'A sudden increase of wind speed by at least 16 knots (8 m/s), the speed rising to 22 knots (11 m/s) or more and lasting for at least one minute. *Note*: when the Beaufort scale is used for estimating wind speed, the following criteria should be used for the reporting of squalls: a sudden increase of wind speed by at least three stages of the Beaufort scale, the speed rising to Force 6 or more and lasting for at least one minute.'

5.1.2.3. *Gust*:   a rapid increase in the strength of the wind relative to the mean strength at the time. A gust is of shorter duration than a squall and is followed by a 'lull' or slackening of the wind speed.

5.1.2.4. *Veering*:   a clockwise change in wind direction; for example, from south to west through south-west.

5.1.2.5. *Backing*:   a counter-clockwise change in wind direction; for example, from south to east through south-east.

5.2. EXPOSURE OF SURFACE WIND SENSORS

The motion of the air near the earth's surface is much affected by such factors as the roughness of the ground, the type of surface, heat sources, the presence of buildings, trees etc.; moreover, wind speed normally increases with height above the earth's surface. It is therefore necessary to specify a standard exposure for making measurements of surface wind so that observations made at different locations may be compared. As mentioned in 5.1, the standard height for surface wind measurements is 10 metres above the ground and they should be made over open and level terrain. Where there are surrounding obstacles such as buildings or trees disturbing the flow of air, it is necessary to increase the height in order to obtain an exposure which is virtually clear of these disturbances but which is as nearly as possible equivalent to that at 10 metres over open level ground nearby. Notes on the siting of anemometers are given in section I.12 of Appendix I.

5.2.1. Effective height.   Each anemometer is allotted an 'effective height' which is defined as the height above open level terrain in the vicinity at which mean wind speeds would be the same as those actually recorded by the anemometer. For example, if in order to minimize the effect of nearby buildings an anemometer is exposed at a height of 20 metres, but is believed to record mean wind speeds equal to those which would occur at a height of 8 metres in an open situation nearby (if such existed), then the effective height is said to be 8 metres.

It is evident from this definition that the determination of effective heights is, in general, a subjective process and no definite rules can be stated. At each station with an anemometer, account will be taken of the nature, extent, height and distance of any obstacles to the wind flow and of the actual height of the anemometer itself. The assessment of the effective height is then a matter of judgement; in some more extreme cases this may result in the effective height having a different value according to the wind direction. The final figure for a station is notified by the Climatological Services Branch of the Meteorological Office after consideration of all the factors.

5.2.2. Correction of wind to standard height at land stations. When anemometers have to be installed at effective heights differing substantially from 10 metres, neither mean speeds nor gusts are strictly comparable with those recorded at sites with a standard exposure. Tabulations made for climatological purposes will record speeds actually measured; appropriate corrections will be made at Headquarters when the data are fully analysed and before they are used, for example, to answer enquiries. When making reports for synoptic purposes, gust speeds are similarly sent without corrections, but the following corrections should be applied to all values of 10-minute mean wind speeds before these are encoded.

Effective height metres	Correction
1–2	add 30 per cent
3–4	add 20 per cent
5–7	add 10 per cent
8–13	no correction
14–22	subtract 10 per cent
23–42	subtract 20 per cent
43–93	subtract 30 per cent

The corrections above are based on the following formula:

$$V_h/V_{10} = 0 \cdot 233 + 0 \cdot 656 \log_{10}(h + 4 \cdot 75)$$

where V_h = speed (in knots) at height h

V_{10} = speed at 10 metres

h = height in metres

from which the following ratios are calculated:

Height in metres	1	2	3	4	5	10	15	20	25	30	40	50
Ratio of speed to that at 10 m	0·73	0·78	0·82	0·85	0·88	1·00	1·08	1·15	1·20	1·24	1·32	1·37
Reciprocal	1·37	1·28	1·22	1·18	1·14	1·00	0·93	0·87	0·83	0·81	0·76	0·73

Investigations have indicated that the above values are sufficiently accurate for ordinary use at land stations. Strictly speaking, the rate of change of wind speed with height varies with the lapse rate of temperature (thermal correction), with wind speed (the extent of mechanical turbulent mixing) and with terrain (surface friction and topographically induced eddies), but the observer is not generally required to take account of all of these effects in applying routine corrections. Note that the corrections above should not be applied to wind speeds observed over the sea.

5.2.3. Correction of wind to standard height over the sea. Wind measurements over the sea, for example at data-collecting platforms, are usually made at heights a long way from the standard 10 metres. The principles for correction are the same as those for land stations: corrections are applied at the station only to reports of 10-minute mean wind speeds made for synoptic purposes. In a marine environment, however, the value of the correction to be applied differs significantly from that over land and is more appropriately based on the power law

$$V_h/V_{10} = (h/10)^b$$

where V_h is the mean wind speed (over 10 minutes) at a height of h metres,

V_{10} is the mean wind speed (over 10 minutes) at a height of 10 metres,

h is the effective height of the anemometer above mean sea level.

Following the agreement reached by the North Sea Meteorological Panel, the value for the index b has been taken as 0.13. The following ratios are calculated from this power law:

Height in metres	10	20	30	40	50	60	70	80	90	100
Ratio of speed to that at 10 m	1·00	1·10	1·15	1·20	1·23	1·26	1·29	1·31	1·33	1·35
Reciprocal	1·00	0·91	0·87	0·83	0·81	0·79	0·78	0·76	0·75	0·74

The following corrections should therefore be made.

Effective height *metres*	Correction
10–13	no correction
14–35	subtract 10 per cent
36–90	subtract 20 per cent.

5.3. ESTIMATION OF WIND DIRECTION AND SPEED

In the absence of instruments, or when instruments become unserviceable, it is necessary to estimate the wind direction and speed. Even at stations which have instruments, observers should make a practice of estimating the wind direction and wind force when outside at observation times. Not only does this become a check against possible instrumental errors, but it also develops a skill which will be valuable if the instruments become unserviceable.

In estimating wind direction and force the observer should stand, if possible, in an open situation avoiding the vicinity of buildings, trees or any similar obstruction, because even small obstructions may cause a significant change in wind speed and deviations in wind direction, especially on their lee side.

5.3.1. Estimation of direction by visual observation. A wind vane specially designed for the purpose is the best device for obtaining the wind direction for meteorological purposes. The requirements are:

(a) It must be sensitive so that it correctly indicates the direction of all but the very lightest winds.

(b) It must be accurately balanced so that it has no tendency to set in a particular direction.

(c) It must be freely exposed and high enough above buildings and trees not to be affected by eddies created by them.

(d) It should be furnished with fixed arms indicating the true directions of the cardinal points and be so placed that the observer can stand nearly under it.

At stations maintained by the Meteorological Office the wind vane supplied for direct visual readings of the wind direction is the standard Meteorological Office wind vane Mk 2B, installed in such a way as to comply with requirements (c) and (d) above. The prime requisite for the mast on which such a vane is mounted is that it should be rigid and vertical and preferably without the impediment of guy wires which constitute a hazard. The vane should not be so high as to be difficult to see from the ground at night. Each site must be judged on its merits and some compromise reached between the often conflicting requirements of exposure, safety and visibility of the vane. Ideally the vane should be installed in an open position on a steel mast approximately 6 metres high. Where the situation is obstructed by trees etc., the vane may be erected on a building or high mast so that it is higher, by at least 3 metres, than the highest obstacle in the immediate vicinity; (see also I.11 (page 196) in Appendix I).

Wind vanes installed on church spires and public buildings rarely meet all the requirements stated above and should not be used. The direction of movement of clouds, however low, should be ignored because wind direction normally changes with height. Visual observation of a drogue or wind-sleeve, a light streamer attached to a suitable mast, or the drift of smoke from low chimneys may be helpful; but care and experience are necessary to eliminate errors due to perspective when the observer cannot stand directly below the indicator.

When the visual observation of vanes or other low-level indicators cannot be made the observer should determine the wind direction by standing in the most exposed part of the observation area and facing into wind; the direction from which the wind is coming may then be determined by reference to the true direction of known landmarks. However, care should always be taken to guard against mistaking local eddies due to buildings, trees, etc. for the general drift of the wind.

5.3.2. Estimation of wind speed. At stations not equipped with an anemometer the strength of the wind must be estimated. Estimates are based on the effect of the wind on movable objects and on the observer's own sensations. The effect of buildings and trees on the mean wind has been mentioned above. When estimating wind speed the observer should stand on flat open ground, well away from such obstructions. Wind speed will be particularly affected on their lee side.

A convenient scale for estimating wind strength is the Beaufort scale of wind speed, so named after its originator Admiral Sir Francis Beaufort. The specifications of the scale now differ from those originally devised by him in 1806 but the present scale has continued in use, substantially unchanged, for many years. The observer should first estimate the force on the Beaufort

BEAUFORT SCALE: SPECIFICATIONS

Force	Description	Specifications for use on land	Specifications for use at sea
0	Calm	Calm; smoke rises vertically.	Sea like a mirror.
1	Light air	Direction of wind shown by smoke drift, but not by wind vanes.	Ripples with the appearance of scales are formed, but without foam crests.
2	Light breeze	Wind felt on face; leaves rustle; ordinary vanes moved by wind.	Small wavelets, still short but more pronounced. Crests have a glassy appearance and do not break.
3	Gentle breeze	Leaves and small twigs in constant motion; wind extends light flag.	Large wavelets. Crests begin to break. Foam of glassy appearance. Perhaps scattered white horses.
4	Moderate breeze	Raises dust and loose paper; small branches are moved.	Small waves, becoming longer; fairly frequent white horses.
5	Fresh breeze	Small trees in leaf begin to sway; crested wavelets form on inland waters.	Moderate waves, taking a more pronounced long form; many white horses are formed. Chance of some spray.
6	Strong breeze	Large branches in motion; whistling heard in telegraph wires; umbrellas used with difficulty.	Large waves begin to form; the white foam crests are more extensive everywhere. Probably some spray.
7	Near gale	Whole trees in motion; inconvenience felt when walking against wind.	Sea heaps up and white foam from breaking waves begins to be blown in streaks along the direction of the wind.
8	Gale	Breaks twigs off trees; generally impedes progress.	Moderately high waves of greater length; edges of crests begin to break into spindrift. The foam is blown in well-marked streaks along the direction of the wind.
9	Strong gale	Slight structural damage occurs (chimney-pots and slates removed).	High waves. Dense streaks of foam along the direction of the wind. Crests of waves begin to topple, tumble and roll over. Spray may affect visibility.
10	Storm	Seldom experienced inland; trees uprooted; considerable structural damage occurs.	Very high waves with long overhanging crests. The resulting foam, in great patches, is blown in dense white streaks along the direction of the wind. On the whole the surface of the sea takes a white appearance. The 'tumbling' of the sea becomes heavy and shock-like. Visibility affected.
11	Violent storm	Very rarely experienced; accompanied by widespread damage.	Exceptionally high waves (small and medium-size ships might be for a time lost to view behind the waves). The sea is completely covered with long white patches of foam lying along the direction of the wind. Everywhere the edges of the wave crests are blown into froth. Visibility affected.
12	Hurricane	—	The air is filled with foam and spray. Sea completely white with driving spray; visibility very seriously affected.

AND EQUIVALENT SPEEDS

Force	Specifications for coastal use	Equivalent speed at 10 m above ground					
		Knots		Miles per hour		Metres per second	
		Mean	Limits	Mean	Limits	Mean	Limits
0	Calm.	0	<1	0	<1	0·0	0·0–0·2
1	Fishing smack* just has steerage way.	2	1–3	2	1–3	0·8	0·3–1·5
2	Wind fills the sails of smacks which then travel at about 1–2 knots.	5	4–6	5	4–7	2·4	1·6–3·3
3	Smacks begin to careen and travel at about 3–4 knots.	9	7–10	10	8–12	4·3	3·4–5·4
4	Good working breeze, smacks carry all canvas with good list.	13	11–16	15	13–18	6·7	5·5–7·9
5	Smacks shorten sail.	19	17–21	21	19–24	9·3	8·0–10·7
6	Smacks have double reef in mainsail. Care required when fishing.	24	22–27	28	25–31	12·3	10·8–13·8
7	Smacks remain in harbour and those at sea lie-to.	30	28–33	35	32–38	15·5	13·9–17·1
8	All smacks make for harbour, if near.	37	34–40	42	39–46	18·9	17·2–20·7
9	—	44	41–47	50	47–54	22·6	20·8–24·4
10	—	52	48–55	59	55–63	26·4	24·5–28·4
11	—	60	56–63	68	64–72	30·5	28·5–32·6
12	—	—	≥64	—	≥73	—	≥32·7

*The fishing smack in this table may be taken as representing a trawler of average type and trim. For larger or smaller boats and for special circumstances allowance must be made.

scale, making use of the criteria set out on page 86, and record the equivalent mean wind speed in knots in the Register. It should be emphasized that the criteria for use on land are not intended to be in any way precise, but are merely to indicate the kind of effect produced by different forces. Force 8 is uncommon inland and represents a wind which is unmistakably 'blowing a gale' and difficult to walk against. The (archaic) coastal specifications (page 87) were based on the behaviour of a fishing smack and may be of assistance to observers on the coast who are knowledgeable in the management of small sailing craft. Where the station commands a good view of the open sea the specifications for use at sea are applicable, but they should not be applied to harbours or other enclosed waters.

Comparisons between estimates of force (B) on the Beaufort scale and measurements of the speed (V) of the wind at an effective height of 10 metres above the ground have led to the establishment of the following relationship between B and V expressed in various units:

$V = 1.87 \sqrt{B^3}$ when V is measured in miles per hour.

$V = 1.625 \sqrt{B^3}$ when V is measured in knots.

$V = 0.836 \sqrt{B^3}$ when V is measured in metres per second.

A table based on these relationships is given on page 87.

5.4. WIND SPEED MEASUREMENT

Instruments used for measuring the wind speed are called anemometers. In the United Kingdom horizontal wind speed is reported in knots (kn) for both climatological and synoptic purposes. Other units such as metres per second, feet per second, miles per hour and kilometres per hour are, or have been, used in certain circumstances and the relations between them are shown in the following table.

m/s	kn	mile/h	ft/s	km/h
1	1·944	2·237	3·281	3·600
0·514	1	1·151	1·688	1·852
0·447	0·869	1	1·467	1·609
0·305	0·592	0·682	1	1·097
0·278	0·540	0·621	0·911	1

Some instruments do not respond to wind speeds of less than 5 knots (2·5 m/s) and so speeds of less than 5 knots must often be estimated.

5.4.1. Cup anemometers.   The standard instruments used by the Meteorological Office for measuring wind speed are called cup anemometers. Three or more cups, roughly conical in cross-section, are mounted symmetrically on arms set at right angles into a vertical spindle. The wind blowing into the cups causes the spindle to rotate, and in standard instruments the design of the cups is such that the rate of rotation is directly proportional to the speed of the wind to a sufficiently close approximation.

The rotation of the cups can be utilized in several ways to obtain values of wind speed. The various cup anemometers used by the Meteorological Office are described below.

5.4.1.1. *Counter anemometers.* The Meteorological Office anemometer, cup counter, Mk 2, is designed primarily for measuring the run of wind over a period of hours or a whole day, rather than over the short period required for synoptic purposes. The short spindle to which the three conical cups are attached is connected by worm-gearing to a revolution counter. The gear ratio is so chosen that the counter indicates directly in miles, tenths and hundredths. Some instruments have the counter window in a vertical position and others have the counter window angled to facilitate reading from the ground.

This instrument is used at many agrometeorological stations to record the run of wind at 2 metres above ground level, an observation used, for example, in the estimation of potential evapotranspiration.

5.4.1.2. *Contact anemometers.* These anemometers are fitted with a switch mechanism which makes an electrical contact at a frequency proportional to the wind speed.

The Meteorological Office cup generator/contact anemometer, Mk 4A, in addition to the generator (see 5.4.1.4), has a micro-switch which is actuated by a falling weight. A worm-wheel from the cup spindle operates on a cam which is raised by the action of the worm-wheel and then released. In falling, it operates the micro-switch. The 'on' period of the switch is therefore independent of the wind speed and the switch cannot remain in the 'on' position. In this way the possibility of the switch remaining in the 'on' position during a calm, or of very long contacts in a light wind, is avoided. The micro-switch is actuated once every 49 revolutions of the cup spindle, this being equivalent to two contacts in 10 minutes per knot of mean wind speed.

The contacts can either be counted by means of a remote audio signal or registered on an electromagnetic counter unit.

5.4.1.3. *Hand anemometers.* The Meteorological Office hand anemometer (see Plate XVII) is based on the magnetic-drag principle and is used to obtain values of wind speed at the level of the observer. The instrument consists of three or four small cups mounted on a vertical spindle. The mechanism and indicating scale are mounted in a cylindrical housing below the cups, and below the housing is a short handle by which the instrument is held. The indicating scale is calibrated in knots over a range of 0 to 60. Each instrument is supplied with a calibration card, and instruments supplied by the Meteorological Office should be returned for recalibration at regular intervals of about one year. Any instrument corrections should be applied before effective height corrections.

In use, the anemometer is held with its axis vertical at arm's length and with the arm at right angles to the wind direction to ensure that the disturbance of the airflow, caused by the observer's body, is reduced as much as possible.

At least two readings of mean wind speed (each reading being taken over at least 15 seconds) are taken within the overall period of observation. When the readings differ by less than 10 knots their average is reported. When the readings differ by 10 knots or more a third reading is taken. If this third reading is within 10 knots of the first reading then the average of all three readings is reported. If the third reading differs by 10 knots or more from the first reading, but is within 10 knots of the second reading, then the change is taken as real and the average of the last two readings alone is reported. Additional mean values are taken, if necessary, to fulfil these conditions.

When the measured wind speed is required for synoptic reports, the speed must be corrected to the standard height of 10 metres (see 5.2.2, or 5.2.3 if over the sea). If the observer stands at ground level the speed reading has to be increased by 30 per cent.

The hand anemometer must be treated with great care; rough handling and vibration may cause demagnetization, resulting in a change of calibration. To protect it from damage it must be stowed in its box when not in use and it must not be put down where it is liable to be affected by stray magnetic fields, especially alternating ones, caused by adjacent electrical apparatus such as motors, relays, transformers, etc. The cups are particularly liable to damage just before, or after, an observation when the cups may be rotating at speed. If they inadvertently strike an object at this stage then the damage can be severe. Even minor damage will change the calibration.

5.4.1.4. *Generator anemometers.*   The rotation of the cups in this pattern of anemometer is used to drive an electrical generator; the voltage generated increases with the speed of the wind and this voltage is indicated on a dial graduated directly in knots. Cup generator anemometers are listed below.

(*a*) Anemometer, cup generator, Mk 2, suitable for operating a recorder and up to six indicating dials connected in parallel. It may be fitted with a strengthened cup assembly capable of withstanding speeds up to 180 knots. When mounted above a Mk 3B wind vane the 'in-line' assembly forms the head of the Meteorological Office electrical anemograph Mk 2.

(*b*) Anemometer, cup generator/contact, Mk 4A (see 1 in Plate XVIII), suitable for operating a recorder and up to six indicating dials connected in parallel. It will withstand wind speeds up to 180 knots. When mounted above a Mk 4G wind vane the 'in-line' assembly forms the head of the Meteorological Office electrical anemograph Mk 4A.

(*c*) Anemometer, cup generator, Mk 5, is externally identical to the Mk 4 but has a starting speed of about $2\frac{1}{2}$ knots due to the use of a generator with lower magnetic drag, and is described in 5.7 (page 94).

The types described in (*a*) and (*b*) can both be used independently of their respective wind vanes.

5.4.2. Readings from anemometers.   At a station equipped only with an indicating dial anemometer the observer should take at least two readings of mean wind speed (each reading being averaged over at least 15 seconds) within the overall period of the observation and should report the average of these readings. When these readings differ by 10 knots or more, a further 15-second mean reading should be taken. If this third reading agrees with the first reading, within a limit of 10 knots, the mean of all three values should be reported. If the third reading does not agree with the first but agrees with the second, the change should be taken as real and the mean of only the last two readings is reported. Additional mean values are taken, if necessary, to provide a set of two or three successive readings that fulfil these conditions.

5.4.3. Sudden change of wind speed.   Particulars of any sudden change of wind speed should be noted in the remarks column of the Register.

5.5. WIND DIRECTION MEASUREMENT

Continuous remote indication of wind direction is provided by means of synchronous transmitters and receivers. The transmitting unit is mounted directly above the vane and its rotor is connected through gearing to the vane spindle. The movements of the vane are transmitted to the receivers and thence to the dial pointers. Either a 12-volt d.c. Desynn system or a 50-volt a.c. magslip system can be employed. The wind vanes used by the Meteorological Office are listed below.

(a) Wind vane, remote-transmitting, Mk 3B, which drives, through gearing, a magslip transmitter. Above the vane and transmitter is a platform to which the base of a Mk 2 anemometer may be bolted so that the vane and the anemometer form a single transmitting head for the Meteorological Office electrical anemograph, Mk 2.

(b) Wind vane, remote-transmitting, Mk 4A, which drives, through gear wheels and a countershaft, a Desynn transmitter. At the top of the casing containing the Desynn is a platform to which the base of a Mk 4A anemometer may be fastened so that the vane and anemometer form a single transmitting head. Alternatively, if the vane is used by itself, a cap is fitted to the top in place of the anemometer.

(c) Wind vane, remote-transmitting, Mk 4G (illustrated in Plate XVIII as forming part of the Meteorological Office Mk 4 wind system), is similar to the Mk 4A described in (b), but a magslip transmitter is employed.

5.5.1. Reading the direction indicator. At a station equipped with a wind vane which has remote-indicating dials, the observer should take at least two readings of mean wind direction (each reading being a mean over at least 15 seconds) within the overall period of the observation and should report the average of these readings. When the first two readings differ by 30 degrees or more, a third final 15-second mean reading should be taken. If this final reading agrees with the first reading, within the limit of 30 degrees, the mean of all three values should be reported. If the final reading does not agree with the first one but agrees with the second, the change may be taken as real and the mean of only the last two readings is reported. It is possible that on occasions of light air (Beaufort force 1) all three readings will differ by 30 degrees or more. In that case the last reading will be reported.

5.5.2. Sudden change of wind direction. Particulars of any sudden change of wind direction should be noted in the remarks column of the Register, but discretion is needed on occasions of very light or no wind when the direction indicator may vary over a wide range of directions in a comparatively short space of time.

5.6. ANEMOGRAPHS

Strictly speaking, an anemograph is an instrument for recording the speed of the wind but, by convention, the term is used in the United Kingdom to cover direction also. The standard instrument used by the Meteorological Office is the electrical anemograph which records the wind speed from the output of a cup generator anemometer and the wind direction from a remote-transmitting wind vane.

The Meteorological Office electrical anemograph, Mk 4, is illustrated in Plate XVIII. The anemograph recorder and dials may be sited at a distance of approximately 500 metres from the transmitting head, depending on the conductor resistance of the cable link. Six indicating dials for speed and direction may be connected in parallel with the recorder.

5.6.1. Anemograph recorder.   The recorder consists of speed and direction units mounted side by side with pens recording on a dual-scaled double-width chart (see 5 in Plate XVIII). The chart mechanism is driven by a synchronous motor which has a standard speed of 1 inch per hour and the chart is capable of taking a continuous record for 31 days.

The recorder is fitted with a range-change switch by which the range can be changed from 0–90 knots to 0–180 knots. The switch is usually kept in the 0–90 position, but when a wind speed of 70 knots is first recorded the switch is moved to the 0–180 knot position. When the mean wind speed falls below 50 knots the switch is reset to the normal range. Recorders at certain stations have the manual range-changing switch removed and an additional unit coupled to the recorder which will automatically switch to the longer range when the wind speed exceeds 70 knots. A red indicator lamp is illuminated when this occurs. Once switched over, the recorder will remain on the 0–180 knot range until the 0–90 knot reset button is pressed. This should not be done until the mean wind speed falls below 50 knots. See 5.6.3 for instructions on noting the times in the Register and indicating on the chart the duration of the change.

Plate XIX shows a section of a typical trace from an electrical anemograph. The upper trace is the record of wind direction and the lower that of wind speed. The direction trace illustrates the effects of turbulence which causes short-period fluctuations of the wind, and therefore of the pen. The result is a broadening of the trace to produce a typical spread of 40 or 50 degrees (or even more at some stations). The lower record of wind speed is also broadened by fluctuations and shows gusts and lulls. The average difference between the speeds of the gusts and lulls is usually a substantial percentage of the mean speed. This percentage, known as the 'gustiness factor' varies with the nature of the terrain in the vicinity of the anemometer. For British stations it is likely to be in the range 25 to 100 per cent. It is smallest over the open sea and greatest in urban or woodland sites because of eddies in the wind stream set up by obstacles to the surface flow. To illustrate the variation in the 'mean wind' a white line has been drawn over the speed record. At open unobstructed sites, and with no neighbouring topographical features that would tend to produce larger-scale eddying flows (mainly in the vertical), the line of the mean wind on the record may usually be assumed to lie midway between imaginary lines drawn through the peaks of the gusts and the troughs of the lulls. This is not true, however, for obstructed sites where the gustiness is greater and eddying circulations will be induced both in the horizontal and the vertical. In such cases the gusts are likely to rise higher above the mean line than the lulls fall below it, since the effect of the obstructions will at times be to direct the stronger wind flow at higher levels down towards the surface. Where there are circulations generated by topographical features, such as those in the lee of hills, the topographically induced eddy may at times oppose the flow of the free wind. The uncritical use of the observed lulls to arrive at the mean free wind speed would then lead to an underestimate. Similarly it would also be possible to choose other, not too distant, observing points for the wind, where the air motions generated by the ground-surface contours commonly lead to an overestimate of the free wind.

These remarks are intended to emphasize the general need for caution in the selection of observing sites and in the interpretation of the records obtained.

5.6.2. Readings from anemographs.   At a synoptic station equipped with an anemograph the observer reports the mean wind speed and mean wind direction recorded over the last 10 minutes. However, when there has been a change of 10 knots or more in mean wind speed and/or 30 degrees or more in the mean direction, and this has been maintained for at least 3 minutes, the observer reports the new speed and/or direction. For synoptic reports the mean wind speed should be corrected, if necessary, for effective height.

At a synoptic station there is also a requirement for reporting additional information about times of onset and cessation of gales, extreme speed and gusts etc. Details of requirements and reporting procedures are to be found in the *Handbook of weather messages,* Part III.

5.6.3. Time marks and notations on the chart.   When the range-change switch (see 5.6.1) is in use and the recorder is on the 0–180 knot range, it is most important that a note should be written on the chart giving the time of the change-over; another note should be written on the chart when the change back to normal response is made, and a firm line, of a distinctive colour, should be drawn above the speed trace throughout the duration of the reduced chart scale.

Daily time marks can best be made by using a soft pencil or ball-point pen to make a short stroke or legible dot at a place on the speed chart, in line with the pen, not covered by the trace. The exact time should be noted in the Daily Register, where one is kept, or in a separate notebook. Dates and times should be written against the marks on the chart either at the same time or subsequently after the chart roll has been removed from the recorder.

5.6.4. Care of the recorder.   The recorder should need little attention other than keeping it clear of dust and spilt ink, topping up the ink-wells and changing the chart as necessary. When topping up the ink-wells, care should be taken to avoid both dislodging the pen from the pen stirrup and overfilling the ink reservoir. Excess ink can damage the pen mechanisms and even cause seizing of the speed movement. A small kit of accessories is supplied with each anemograph which allows for the cleaning or replacement of the pens.

About once a month check that all speed dials, where fitted, agree with each other and with the recorder to within 2 knots below 40 knots and to within 4 knots above that value. Direction dials, if fitted, should agree with each other and the recorder to within 6 degrees. All pointers and pens should be moving freely and smoothly at all times. During flat calms or when the recorder is switched off for any purpose, check that the zero of the speed pen lies accurately on the zero of the chart. Any significant faults shown up by the above checks should be reported to the appropriate authority.

5.7. THE METEOROLOGICAL OFFICE Mk 5B WIND SYSTEM

In the electrical anemograph, Mk 4 (5.6 refers), the voltage generated by the anemometer passes directly through cables to the dials and chart recorder. In this system the calibration is based on having a fixed resistance value for the circuit and this is provided by the dial and recorder and by fixed resistors which may be removed as dials are added. However, this system is unsuitable if many additional displays are required or where the length of the cable between the anemometer tower and the displays has too high a resistance. In such circumstances the Mk 5B wind system may be used. The heart of the system is the Mk 5B converter unit which takes the outputs from a Mk 4G wind vane and either a Mk 4A or Mk 5 cup generator anemometer.

The wind-vane input data are converted into d.c. voltages over the range 0–10 volts, representing a range of 540 degrees. Logic circuits shift the scale reading by 360 degrees whenever the extreme values of 0 and 540 degrees are approached. This avoids any discontinuities when the wind vane passes through north.

The cup generator anemometers produce voltages, the amplitude and frequency of which increase or decrease according to the wind speed. A converter is used to convert the frequency to a d.c. voltage in the range 0–10 volts, representing 0–200 knots.

The sensors can be linked by cable to the Mk 5B converter unit over distances up to 3 km with further cable links of about 500 metres to the Mk 5C wind speed and direction meters, and 100 metres for the Mk 5B recorders. For distances greater than 3 km a Mk 5B telemetry system, operating over telephone lines, has to be employed.

The Mk 5B recorder (see Plate XX) produces a record very similar to that of the Mk 4G. The chart drive speed is 30 mm per hour. A control unit may be added which will automatically change the range of the recorder to 0–200 knots when the wind speed exceeds 70 knots. A warning light is illuminated when this occurs. The recorder will remain on the extended range until manually reset.

5.8. OTHER WIND SYSTEMS

In addition to the wind systems and sensors described above, a number of other systems are employed in the Meteorological Office. Some are characterized by smaller size and lower inertia of the cup rotors and vanes and generally employ photoelectric devices to detect motion. They are usually associated with automatic weather stations of various types and in all cases their use will have had prior approval from Meteorological Office Headquarters.

CHAPTER 6

STATE OF GROUND AND CONCRETE SLAB

6.1. GENERAL

Observations of the state of ground and of a concrete slab form part of the normal routine at synoptic and some climatological stations. At synoptic stations, state-of-ground observations are made at each synoptic hour (that is three-hourly from 0000 GMT) and state-of-concrete-slab observations are made at 0900 GMT. At climatological stations, observations of state of both ground and concrete slab are recorded once a day, at 0900 GMT, and, exceptionally, state of ground at other hours laid down for the station. These observations are included in reports only at the times detailed in the *Handbook of weather messages,* Parts II and III, and *Abbreviated weather reports*.

The state of the concrete slab is selected from the scale given in 6.3.

For the state of ground, the observer selects the state from one of the scales given in 6.2.1 and 6.2.2. Some of the states are defined by reference to representative bare ground which is usually a 2-metre square in the meteorological enclosure, as described in 6.1.1. The other states are defined with reference to the open area representative of the station, as described in 6.1.2.

It should be noted that the new state-of-ground codes, introduced by WMO with effect from 1 January 1982, allows somewhat greater detail than previously in reporting the state of ground. This is done by allocating two sets of code figures 0 to 9, one set for the state of ground WITHOUT snow or measurable ice cover and the other set for state of ground WITH snow or measurable ice cover (see 6.2.1 and 6.2.2).

6.1.1. Representative bare ground should be a bare plot about 2 metres square with its surface level with the surrounding ground and preferably in the meteorological enclosure (as indicated in Figure 20 on page 179) but, if necessary, sited conveniently elsewhere. The bare plot must not be on top of a mound, on a slope, in a hollow, in the neighbourhood of a spring, or in other locations not truly representative of the station site. At least the top 15 centimetres of the plot should be soil which is a fair sample of the soil in the locality; it should be kept free from weeds, if necessary by the use of weed-killer, but otherwise disturbed as little as possible and allowed to consolidate naturally.

6.1.2. Open area representative of the station should be taken to include the open, fairly flat ground easily visible from the station, and not differing from it in altitude by more than 30 metres. Any area with untypical surface characteristics or topography should be excluded from consideration; snow cover and surface moisture, for example, are not likely to be adequately representative on the ground beneath trees, on steep slopes, in a small valley or cutting, or at the bottom of an isolated hollow.

6.2. STATE-OF-GROUND SCALES

The state of ground is selected from one of the following scales. In all inst-ances the highest applicable code figure in either scale is to be reported.

6.2.1. State of ground without snow or measurable ice cover:

0 = Surface of ground dry (without cracks and no appreciable amount of dust or loose sand)

1 = Surface of ground moist

2 = Surface of ground wet (standing water in small or large pools on surface)

3 = Flooded

4 = Surface of ground frozen

5 = Glaze on ground

6 = Loose dry dust or sand not covering ground completely

7 = Thin cover of loose dry dust or sand covering ground completely

8 = Moderate or thick cover of loose dry dust or sand covering ground completely

9 = Extremely dry with cracks.

The definitions in the code for figures 0 to 2, and 4, apply to a representa-tive bare patch, and figures 3, and 5 to 9, to an open representative area.

6.2.2. State of ground with snow or measurable ice cover:

0 = Ground predominantly covered by ice

1 = Compact or wet snow (with or without ice) covering less than one-half of the ground

2 = Compact or wet snow (with or without ice) covering at least one-half of the ground, but ground not completely covered

3 = Even layer of compact or wet snow covering ground completely

4 = Uneven layer of compact or wet snow covering ground completely

5 = Loose dry snow covering less than one-half of the ground

6 = Loose dry snow covering at least one-half of the ground (but not completely)

7 = Even layer of loose dry snow covering ground completely

8 = Uneven layer of loose dry snow covering ground completely

9 = Snow covering ground completely; deep drifts.

The definitions in this code apply to an open representative area. Whenever reference is made to ice it also includes solid precipitation other than snow.

6.2.3. Hoar frost and dew.
The scales for the state of ground take no notice of hoar frost or of dew. When these conditions are observed, the state of the ground should be determined independently as if they were absent. They do in fact affect the ground surface, at least temporarily, but are regarded pri-marily as weather observations and are recorded by means of Beaufort letters (see 4.5.1, page 73).

6.3. STATE OF CONCRETE SLAB

The state of the concrete slab is selected from the following scale:

> 0 = Slab dry
>
> 1 = Slab moist
>
> 2 = Slab wet
>
> 3 = Slab icy.

When the concrete slab is covered by snow or the ground is frozen, or the state of the slab cannot be adequately described by the above scale, it is not to be recorded.

CHAPTER 7

ATMOSPHERIC PRESSURE

7.1. GENERAL

The pressure of the atmosphere at any point is the weight of the air which lies vertically above unit area centred at the point.

The instruments most commonly used for measuring the pressure of the atmosphere are the aneroid and mercury barometers. The consistent accuracy required in the measurement of atmospheric pressure for meteorological purposes could, in the past, best be obtained by the use of the mercury barometer. With the development of a precision aneroid barometer of comparable accuracy, the Meteorological Office has adopted the precision aneroid as its standard instrument for this measurement. The precision aneroid is considerably easier to read than the Kew-pattern mercury barometer which it has replaced. It is compact and robust, and its use avoids many of the transportation difficulties of the mercury-in-glass instrument. Stations which report pressure will hold two barometers.

7.1.1. Units of atmospheric pressure. The unit of pressure in the International System (SI) is the newton per metre squared (N/m^2) to which has been given the name pascal and the symbol Pa. The unit for measuring atmospheric pressure for international meteorological purposes, however, remains the millibar (= 100 Pa). The International Committee on Weights and Measures (CIPM) has allotted the symbol mbar to this unit, but the Meteorological Office, together with most other meteorological services, continue to use mb. Barometers issued by the Meteorological Office have their scales graduated in millibars, but older ones may be found with scales graduated in millimetres (mmHg), or inches (inHg), of mercury. The relationships between these pressure units are:

$$1 \text{ mb} = 100 \text{ Pa}$$
$$1 \text{ mmHg} = 133 \cdot 322 \text{ Pa}$$
$$1 \text{ inHg} = 25 \cdot 4 \text{ mmHg.}$$

From these relationships the following table of conversion factors has been prepared, assuming the standard conditions specified in 7.3:

Pressure unit		kPa	mb	mmHg	inHg
1 kPa	=	1	10	7·501	0·295 30
1 mb	=	0·1	1	0·750 1	0·029 530
1 mmHg	=	0·133 32	1·333 2	1	0·039 370
1 inHg	=	3·386 3	33·863	25·40	1

It is probable that at some time in the future the hectopascal (hPa) will be adopted as the unit for measuring atmospheric pressure. This is the exact equivalent of the millibar in the SI system of units. In the meantime either term can be used.

7.1.2. Categories of observations. The following observations relating to atmospheric pressure are included in the schedules of various types of station.

(*a*) Pressure in millibars and tenths, corrected to mean sea level (QFF), at all synoptic stations which report pressure, and at a few climatological stations.

(*b*) Pressure at aerodrome level (QFE), at stations making reports to Air Traffic Control Officers as may be arranged.

(*c*) Altimeter setting (QNH), at stations making reports to Air Traffic Control and for inclusion in aviation weather broadcasts.

(*d*) Barometric tendency and characteristic, at synoptic stations only.

Instructions relating to item (*d*) above, the observation of which requires a barograph, are given in the *Handbook of weather messages*, Part III, to which reference should be made.

7.2. PRECISION ANEROID BAROMETERS

The aneroid barometer consists of one or more shallow capsules of thin metal either completely or partially evacuated according to its design. In older types of aneroids the force on the capsule due to atmospheric pressure was opposed by an external or internal spring, but it is now more common to use a capsule designed to provide its own spring. In this case the material and the corrugated shape of the capsule are chosen as a compromise between the need both to provide a deflexion large enough to be measured simply and reliably and to respond repeatedly to changes in atmospheric pressure. At zero atmospheric pressure the capsule is in its natural unstressed state and as the pressure is increased the capsule is compressed. Typically a change in the atmospheric pressure of 100 mb will produce a deflexion of the capsule of less than 1 mm and this very small movement must be magnified to enable the observer to read the instrument to 0·1 mb.

Until comparatively recently aneroid barometers have not been suitable for measuring absolute values of atmospheric pressure, mainly because of instability in the aneroid capsule resulting in a slow drift of the calibration with time, inadequate compensation for the effect of temperature on the deflexion at a given pressure, and friction and backlash in the system of gears, pivots and levers necessary to provide magnification of the movement for display of the measurement. For these reasons the use of aneroid barometers was for a long time confined to the observation of pressure tendency. Recently, however, the design of aneroid capsules has been improved and better materials have become available. Display systems have also been developed which apply negligible load to the capsule system and eliminate errors due to backlash and friction. Although reduced, some effects remain, arising from long-term drift in the calibration and during transportation; despite these, the aneroid is now preferred to the mercury barometer. The latter is much more difficult to handle in transit and, over a number of years, it can develop errors due to a defective vacuum or fouling of the mercury.

The Meteorological Office uses two versions of precision aneroid barometer, designated Mk 1 and Mk 2 (see Plate XXI). The two differ in appearance but they are identical in their mode of operation. Advice on selecting a

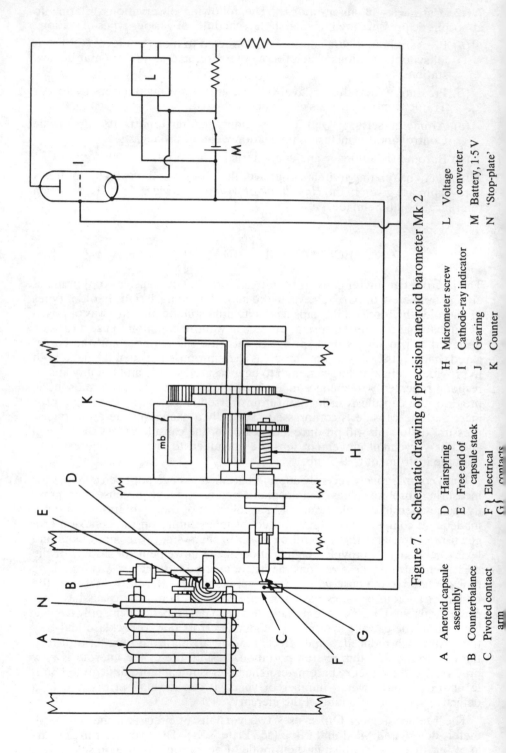

Figure 7.   Schematic drawing of precision aneroid barometer Mk 2

A   Aneroid capsule   D   Hairspring   H   Micrometer screw   L   Voltage
      assembly   E   Free end of   I   Cathode-ray indicator   converter
B   Counterbalance   capsule stack   J   Gearing   M   Battery, 1·5 V
C   Pivoted contact   F } Electrical   K   Counter   N   'Stop-plate'
      arm   G }   contacts

position and on the installation of the precision aneroid barometers is given in Appendix I (para. I.6).

7.2.1. Description of the instrument. Both versions of the instrument have a range from 1050 to 900 mb. The pressure-sensing element is a stack of three aneroid capsules, partially evacuated, and rigidly fixed at one end (A in Figure 7). The other end is free to respond to changes in atmospheric pressure and, in so doing, deflects a pivoted counter-balanced (B) contact arm (C). The pivot is mounted in jewelled bearings. The contact arm is held lightly against the centre of the free end (E) of the capsules by means of a hairspring (D). A contact (F) on the other end of the arm is aligned with another contact (G) at the end of the micrometer screw (H). Movement of the free end of the capsule stack is translated into a displacement of the pivoted arm and this is measured by the use of the micrometer screw. The result is read in a digital counter (K) in millibars and tenths. Contact between the arm and micrometer is sensed electrically and shown by a cathode-ray indicator (I).

The deflexion of the capsule over the pressure range 900 to 1050 mb is approximately 1·27 mm. A stop-plate (N) mounted centrally above the capsule stack prevents expansion to pressures significantly lower than 900 mb. The instrument can therefore be safely transported by air.

7.2.2. Reading the instrument. First switch on the indicator by pressing the black switch button. If the thread of light in the indicator is broken, turn the knurled knob so that the pressure reading decreases, and until the thread becomes continuous. When the light is continuous, gently reverse the movement of the knob so that the pressure reading increases, and until the thread of light just breaks. This dual process is repeated to avoid errors that could arise from 'overshooting' the correct setting on the first attempt. With the thread just broken, the pressure can now be read in the window.

If parts of two figures show equally in the tenth-of-a-millibar position, the odd number is to be taken.

7.2.3. Correction and reduction of readings. The design of precision aneroid barometers includes adequate provision for compensation for changes in temperature, so that readings from the barometer at normal room temperatures do not usually require correction for temperature. As the principle of the instrument involves the balancing of atmospheric pressure against the restoring force of the capsule assembly there is no correction for gravity. Thus, in determining mean-sea-level pressure, the only corrections which need normally be applied are those for the individual calibration of the instrument used and for altitude. The height of the aneroid capsule must be established to an accuracy of 1 metre, if necessary by special survey.

A correction card indicating the calibration errors is issued with, and is unique to, each instrument. The card and instrument should not be separated. On installation, when the final position for the aneroid has been established (and hence its altitude), a further set of corrections will be supplied by the Meteorological Office in order that the readings can be corrected to mean sea level.

7.2.4. Sources of errors in precision aneroid barometers. The precision aneroid barometer may develop a leak in one of its aneroid capsules but this is rare. A more probable defect is a small shift in calibration with time, due to changes in the characteristics of the aneroid capsules. By carrying out routine barometer checks any change in calibration will be detected. If a sudden unexpected change in pressure is indicated by the instrument, the observer should check that a similar change is shown by the spare barometer or by the barograph.

Although the instrument is well compensated for temperature variations, errors can occur if it is exposed to bright sunshine which may give rise to differential heating of the instrument or a general rise of temperature outside the compensating limits. Very rapid temperature changes can also give rise to errors as can exposure in a room in which the temperature goes outside the range of 10 to 30 °C. It follows that an instrument should not be placed too close to artificial sources of heat or in direct sunlight.

7.2.5. The static pressure head. The wind exerts a dynamic pressure proportional to the square of its speed. Eddying flows and wind flows over the surface of buildings generate fluctuating pressures on the building surface and its interior which may be above or below the true atmospheric pressure. Provided ventilation is restricted and is achieved by many small openings well distributed over each face of the room or building (by leakage gaps around windows and doors, for example), as distinct from individually large vents such as open windows or chimneys, then the pressure effects caused by wind are likely to be small. When wind-pressure effects cannot be ignored, and these are more likely to be a problem in tall rather than single-storey buildings, then a static pressure head may be required.

A static pressure head consists of a vertical cylinder with a fin (see Plate XXII); in appearance it is similar to a wind vane and responds to wind direction in the same way. The cylinder is hollow and contains three holes positioned relative to the airflow such that the internal pressure is independent of any dynamic wind effects. A non-moving version in which the vertical cylinder is partially covered by another, conical, cylinder is sometimes encountered. Though the head must be exposed in a similar way to an anemometer, that is as far away as possible from major obstructions to the airflow, the 10-metres height requirement need not be followed. The cylinder is connected to the precision aneroid barometer via a length of plastic tube. It should be noted that, irrespective of the height or location of the static pressure head, the pressure measured is that at the height of the precision aneroid barometer, and a correction for the height of the head above the barometer is not necessary.

A static pressure head may be required when a precision aneroid barometer is installed in an air-conditioned room.

7.3. MERCURY BAROMETERS

The use of mercury barometers for measuring atmospheric pressure has been gradually phased out during recent years. There are a few still in use but if at any time they require replacement they will be exchanged for precision aneroid barometers. A general description of mercury barometers follows so that their mode of operation can be appreciated in the interim period.

In mercury barometers the pressure exerted by the atmosphere is balanced against a column of mercury. Any change in the length of the mercury column is accompanied by a change in the level of the mercury in the cistern. The height of the mercury column depends on atmospheric pressure, the density of the mercury, and gravity. Standard conditions are laid down under which a mercury barometer should read correctly. These are density of mercury at 0 °C, 13 595·1 kilograms per metre cubed (kg/m^3), and a conventional datum for gravity of 9·806 65 metres per second squared (m/s^2).

The height of the mercury column is accurately measured against a fixed scale and an adjustable vernier scale in true units of pressure (millibars). A thermometer attached to the instrument casing (known as the attached thermometer) is used to indicate the temperature of the mercury column from which the density of the mercury can be established. The height of the barometer above mean sea level, the screen dry-bulb temperature and the temperature of the attached thermometer were all factors used in the preparation by the Meteorological Office of a consolidated correction table to which readings made with a mercury barometer are subject if the reduction of the readings to a standard datum level is required.

The sources of error in mercury barometers arise from several factors. The effects of wind around a building can produce fluctuations in the pressure in a room which are superimposed on the static pressure. These can be minimized by careful positioning but the ideal solution of a static pressure head is not easily applicable to mercury barometers. Additionally, the space above the mercury column is assumed to be a vacuum on calibration, but a slow deterioration from this state can cause errors and finally the mercury will in the course of use slowly become contaminated and this also will eventually give rise to errors.

The main difficulty that occurs with the use of mercury barometers is the problem of transporting them.

7.3.1. The Kew-pattern and Fortin barometers.
The Kew-pattern barometer (Figure 8) was the standard issue by the Meteorological Office prior to the introduction of the precision aneroid barometer. In this type the level of the mercury in the cistern does not have to be adjusted as the scale on the barometer is constructed to allow for changes in the level of the mercury cistern.

The Fortin barometer is similar in appearance to the Kew-pattern but there are differences between the two types. The most important is that the level of the mercury in the cistern of the Fortin barometer can be altered so that the surface of the mercury is just in contact with a fixed reference mark, called the 'fiducial point'. This procedure needs to be carried out prior to reading the instrument.

7.3.2. Corrections to mercury barometers.
The procedures for preparing corrections for these barometers are not dealt with in this edition of the handbook. Should the necessity arise for determining them, however, reference will have to be made to the third edition of the *Observer's handbook* (1969) in the case of the Kew-pattern barometer, and to the first edition of the *Handbook of meteorological instruments*, Part I, (1956) in the case of the Fortin. Both editions are available in the National Meteorological Library, Bracknell.

Figure 8.   Kew-pattern barometer Mk 2

7.4. CHECKING BAROMETER READINGS

The Meteorological Office issues detailed instructions concerning the checking of barometer readings at synoptic stations. Routine comparisons are made between the two station barometers in addition to a periodic check comparison with another barometer which is sent to stations specifically for that purpose. For auxiliary stations the check is maintained by the collecting centre, and if at any time doubt arises as to the validity of the readings from a barometer, the auxiliary station will be notified of the appropriate action to take. In addition, during the annual inspection, the station barometers will be further checked against a third, recently calibrated barometer.

At those climatological stations which make pressure readings, a reliable routine check on barometer readings is less easy to achieve. For these stations a comparison can be made periodically between the station barometer readings and those from adjacent synoptic stations. It is therefore recommended that initial contact be made with the Meteorological Office which will suggest a suitable procedure for each station to undertake in order to guard against grossly incorrect readings.

7.5. PRESSURE AT AERODROME LEVEL (QFE)

The calculation of pressure at aerodrome level (QFE) starts from either

(a) the precision aneroid barometer value of pressure at instrument level, i.e. the reading of the precision aneroid barometer corrected for calibration error but not reduced to mean sea level (MSL); or

(b) the mercury barometer reading at cistern level, i.e. the reading of the mercury barometer corrected for index error, temperature and gravity, but not reduced to MSL.

7.5.1. Correction to aerodrome level (QFE). Aerodrome level is defined as the height above MSL of the highest usable point of the landing area. This point is specified for each aerodrome but does not usually differ by more than a few metres from the altitude of the barometer. The value of QFE is therefore obtained by applying a small correction to the pressure at barometer level for the difference in height between the barometer and aerodrome levels. This correction can be obtained from Table I, page 205. When the QFE is set on the altimeter subscale of an aircraft at rest at aerodrome level, the altimeter in the cockpit will show the height of the altimeter above the ground (unless steps have been taken to allow for this in adjustment of the instrument, in which case the altimeter will read zero).

At any station where values of QFE are required, a table should be displayed near the barometer showing the correction for QFE to the nearest 0·1 mb, the corrections being applicable to a range of pressure from 920 to 1040 mb in steps of 20 mb, and a range of temperature from −15 to +35 °C in steps of 5 °C.

It should be noted that QFE corrections are positive when the barometer level is above aerodrome level but are negative when the barometer is below aerodrome level.

7.6. ALTIMETER SETTING (QNH)

There are two main types of altimeter in use in aircraft. These are a radio altimeter or a pressure altimeter. The design and scale graduations of the latter type are based on the concept of a standard atmosphere. The altimeter setting (QNH) is the pressure setting which causes the altimeter to read the height above mean sea level of the touchdown on landing, plus the height of the altimeter above ground (unless an allowance for the latter has already been made in the adjustment of the instrument). The corrections that are applied (see below) to station-level pressure readings are corrections that are appropriate to a standard atmosphere; this ensures that adjustments to an aircraft altimeter are compatible with its design criteria.

At stations where values of QNH are required, a table should be displayed near the barometer showing the corrections required to a range of QFE values. From the height h of the barometer above MSL and the indicated pressure reading corrected in the normal way for index error, obtain a temperature T from the table below. Use this temperature T instead of the dry-bulb temperature in the screen to derive from Table I on page 205 the correction to be applied.

Barometer as read	h (metres)						
	0	50	100	150	200	250	300
mb				*degrees Celsius*			
1040	16·4	16·6	16·8	16·9	17·1	17·3	17·4
1020	15·4	15·5	15·7	15·8	16·0	16·2	16·3
1000	14·3	14·4	14·6	14·8	14·9	15·1	15·3
980	13·2	13·3	13·5	13·7	13·8	14·0	14·2
960	12·1	12·2	12·4	12·5	12·7	12·9	13·0
940	10·9	11·1	11·2	11·4	11·6	11·7	11·9
920	9·8	9·9	10·1	10·2	10·4	10·6	10·7

7.7. BAROGRAPHS

The standard barographs issued by the Meteorological Office are aneroid instruments; their action depends upon the response to variations of atmospheric pressure of disc-shaped capsules made of thin corrugated metal. These capsules are nearly exhausted of air and their surfaces are held apart by an internal spring. If the atmospheric pressure falls, the capsule surfaces move apart. If the pressure rises, the capsule surfaces are compressed and move together. The small movements thus produced in a bank of such capsules are magnified by a system of levers and communicated to a pivot arm that carries a recording pen. The pen is given vertical movement and writes on a chart (barogram) wrapped round a drum which is rotated by clockwork about a vertical axis.

Barographs which employ either mechanical or photographic registration to record the changes of height of a column of mercury have been designed for observatory use. These are generally more accurate than the aneroid type but the latter is more compact, cheaper and quite adequate for routine use.

Plate XVII

METEOROLOGICAL OFFICE HAND ANEMOMETER

Plate XVIII

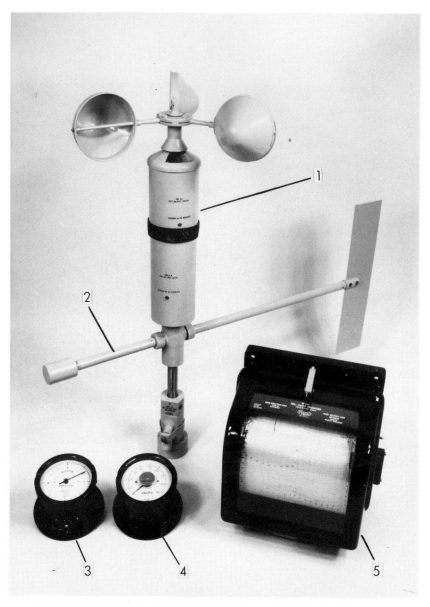

METEOROLOGICAL OFFICE ELECTRICAL ANEMOGRAPH Mk 4

1   Anemometer Mk 4A
2   Wind vane Mk 4G
3   Wind direction indicating dial
4   Wind speed indicating dial
5   Recorder

Plate XIX

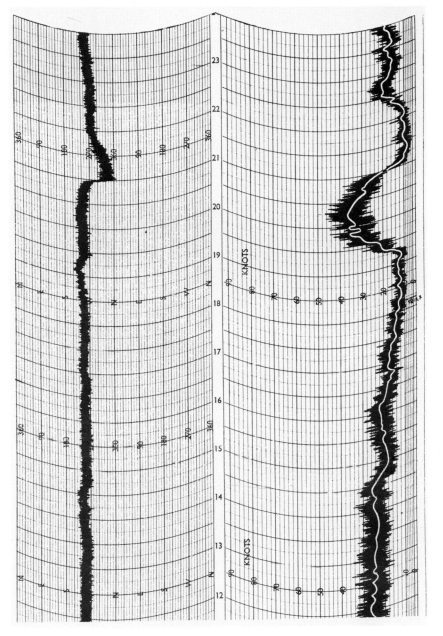

ELECTRICAL ANEMOGRAPH RECORD

Plate XX

METER DISPLAY FOR THE Mk 5 WIND SYSTEM

METEOROLOGICAL OFFICE ANEMOGRAPH RECORDER Mk 5

Plate XXI

PRECISION ANEROID BAROMETER Mk 1

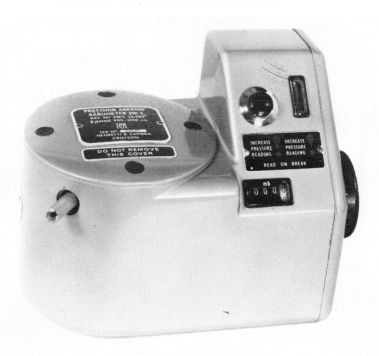

PRECISION ANEROID BAROMETER Mk 2

Plate XXII

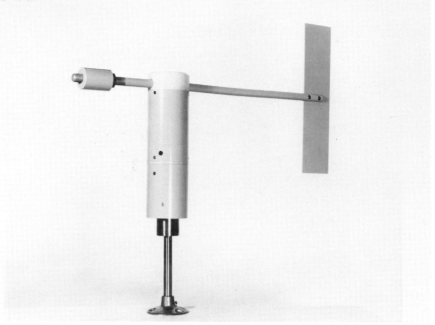

STATIC PRESSURE HEAD

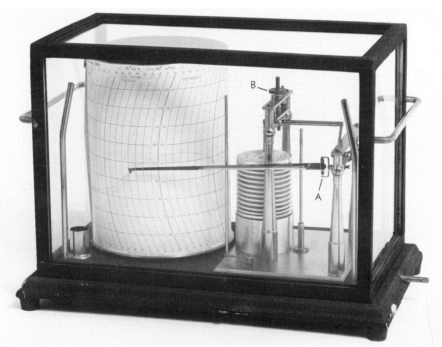

OPEN-SCALE BAROGRAPH

A Gate suspension   B Milled-head adjusting screw

Plate XXIII

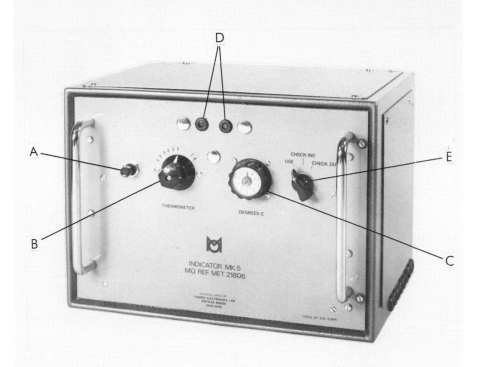

Positive reading +24·6

Negative reading −12·5

TEMPERATURE INDICATOR Mk 5

A On/off switch
B Thermometer selector switch
C Temperature dial (as shown below) and bridge balance switch
D Neon balance indicators
E Calibration switch

Plate XXIV

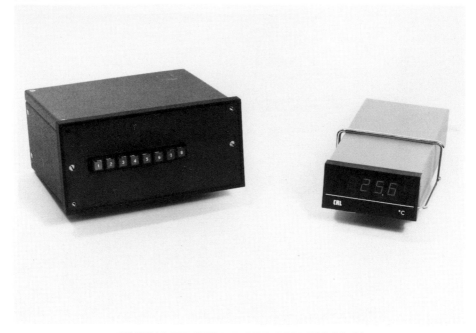

DIGITAL TEMPERATURE INDICATOR Mk 1A

ARRANGEMENT OF INSTRUMENTS
IN A LARGE THERMOMETER SCREEN

Plate XXV

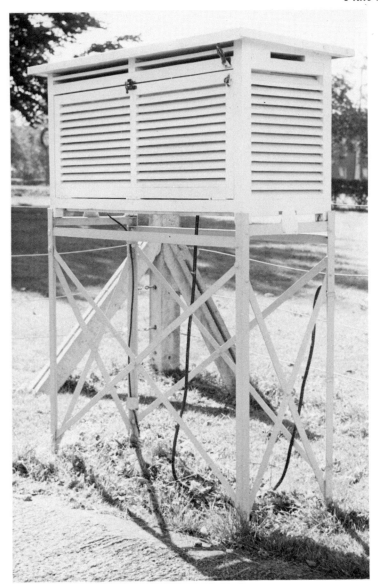

LARGE THERMOMETER SCREEN

Plate XXVI

BIMETALLIC THERMOGRAPH

HAIR HYGROGRAPH

Plate XXVII

ASPIRATED PSYCHROMETER

A   Key for winding clockwork
B   Housing containing clockwork
C   The fan and air outlets
D   Thermometers
E   Main air duct
F   Wet-bulb wick

G   Polished shields
    protecting thermometers
H   Air inlets
I   Rod for supporting
    instrument
J   Point of support when
    instrument is hung
    vertically from the rod

Plate XXVIII

METEOROLOGICAL OFFICE TILTING-SYPHON
RAIN RECORDER Mk 1

Plate XXIX

SUNSHINE RECORDER Mk 2

Plate XXX

SUNSHINE RECORDERS Mk 3C

Plate XXXI

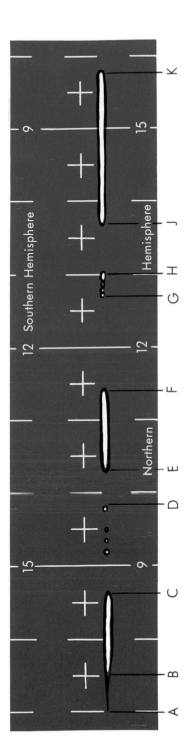

MEASUREMENT OF SUNSHINE CARDS

The illustration shows a typical record on part of an equinoctial card. Marks, indicated by letters A to K, have been made to show between which points the measurements should be made.

The first portion A to B begins faintly after sunrise and slowly increases in width with the intensity of the sun's rays; the measurement is taken from the extreme end of the brown trace. The next portion B to C shows bright sunshine burning completely through the card and the measurement is taken to a point about half-way between the centre of curvature and the extreme visible limit of the burn. From C to D the record consists only of small circular burns; being truly circular these are not measured even when the card is burnt right through. Portions E to F and J to K show continuous bright sunshine and have rounded ends; allowance is made for the spread of the burn by measuring between points about half-way between the centres of curvature and the extreme visible limits of the burns. Finally, portion G to H consists of a series of small circular burns and an elongated burn joined together, no uncharred blue card being visible between them; this is measured in the same way as the continuous burns E to F and J to K.

The hourly amounts are: 07–08 h (Local Apparent Time) 1·0; 08–09 h ·6; 10–11 h ·7; 11–12 h ·4; 12–13 h ·3; 13–14 h ·3; 14–15 h 1·0; 15–16 h ·8; and the total for the day is 5·1 hours.

7.7.1. Exposure of barographs. Meteorological Office barographs are compensated for temperature as far as possible, but exact compensation is difficult to obtain; readings may alter by as much as 0·3 mb for a change of about 5 °C in the temperature of the barograph. The record may also be quite seriously affected if one part of the instrument is at a different temperature from the other parts. It is most important, therefore, that the barograph should not be exposed to direct sunshine, nor to sources of heat which may produce such differential temperature effects. As the barograph is used mainly to obtain the rate of change of barometer readings, sudden changes of temperature should be avoided. The barograph should be in a room where it is not exposed to shaking, vibration or dirt. The room should not be air-conditioned.

7.7.2. Aneroid barograph. Two sizes of aneroid barograph are available, the open-scale barograph (Plate XXII) and the small barograph. They differ somewhat in mechanical details and in the scale of registration but the general principles of design are the same. The recording pen is attached to a very light arm, provided at the far end with a 'gate suspension' (A in Plate XXII). This consists of a two-point bearing tilted slightly inwards at the top so that the line joining the two pivots slopes toward the long axis of the instrument. With this arrangement the pen arm tends to swing inwards under gravity. The pressure of the pen on the paper is thus controlled by gravity and not by the flexibility of the pen arm. It can be adjusted by altering the tilt of the suspension until the pressure is only just sufficient to keep the pen in contact with the chart, and the friction between pen and paper is thus reduced to a minimum.

By means of the milled-head screw B (which is in a different position in some instruments) the position of the pen on the chart can be adjusted. The charts are ruled and figured to cover the range 950 to 1050 mb and the pen should normally be set to indicate the pressure at mean sea level by reference to the corrected readings of either a precision aneroid or mercury barometer. On rare occasions when the barometer reading is exceptionally low or exceptionally high and there appears to be a risk of the pen running off the chart, the pen should be temporarily readjusted by moving it upwards or downwards by, say, 20 whole millibars, so that no part of the record is missed. As soon as pressure returns to more normal values the pen should be brought back to the correct pressure reading. A note should be made in the Register and the chart suitably annotated upon removal.

7.7.3. To change the chart. The clock drum is designed to make a complete rotation in a little over a week and the chart is printed to accommodate a seven-day record starting on Monday. The chart should therefore be changed on Monday mornings at about 0900 GMT, following the procedure detailed below.

(a) Move the pen away from the chart by means of the pen lifter, note the exact time and lift the case of the instrument gently.

(b) Undo the retaining nut in the centre of the drum and lift the drum clear of the clock. Lift the chart-retaining clips and remove the completed chart.

(c) Wind the clock and, if it has been running fast or slow, adjust the regulator.

(*d*) If an inked pen is used, fill the reservoir, first cleaning it if the previous record shows a thick or otherwise unsatisfactory trace. Fit a new pen if necessary. If a fibre-tipped pen is used no routine attention is necessary, the pen being replaced as necessary. Fibre-tipped pens are preferred.

(*e*) Write up the new chart, filling in dates, time of beginning of the record, serial number of the record and details of the station in the spaces provided. Place the chart on the drum, taking care to see that it is in close contact with the drum all round, that the lower edge is touching the flange at the base of the drum, that the registration lines at the two ends of the chart coincide, and that the end of the chart overlaps the beginning and not vice versa.

(*f*) Replace the drum on the clock, taking care to avoid hitting the aneroid assembly, tighten the drum-retaining nut and let the point of the pen nearly touch the chart. Adjust to time by turning the drum backward (counter-clockwise when viewed from above) so as to take up backlash.

(*g*) Close the case of the instrument gently.

(*h*) Using the pen lifter, let the pen point come into contact with the chart. If the pen does not begin to write properly, give it the necessary attention (see 7.7.6).

(*i*) Write up the previous chart, entering the time of ending of the record and, against each daily time mark, the time the mark was made and the pressure at that time, corrected to the datum to which the barograph has been set (usually mean sea level).

The above instructions apply much as they stand to other types of recording instruments such as thermographs and hygrographs with daily or weekly clocks. In some instruments the pen lifter cannot be operated from outside the case.

7.7.4. Time marks.   A time mark should be put on the barogram each day, using the time-marker, if fitted, or by opening the case of the instrument and depressing the pen about 3 or 4 mm. A suitable time may be 1200 GMT. The exact time to the nearest minute should be noted in the Register for subsequent entry on the record itself. If the instrument is free from excessive friction the making of a time mark should not cause a discontinuity, but if there is a small one it should be allowed for, at synoptic stations, in assessing the barometric tendency and characteristic.

7.7.5. Standard for barograph records.   The standard to be aimed for in the record of pressure is a thin, clear and 'lively' trace, showing all the details of the minor changes of pressure. The commonest defects are (*a*) too thick a trace, showing that the pen needs cleaning or renewing; (b) a 'stepped' trace, in which the changes of pressure appear to occur in a series of abrupt jumps, separated by intervals during which the pen traces a horizontal straight line; such a trace affords clear evidence of excessive friction (see 7.7.6); or (c) excessive broadening of the trace in windy weather, due to short-period variations of pressure within the building itself. If this third defect cannot be cured by choosing a different position for the barograph, or by experimenting with the opening or shutting of windows, the substitution by an oil-damped barograph of the type designed for use at sea may be helpful.

7.7.6. Care and attention. A barograph needs little attention beyond changing the charts, daily time marking and inking any metal pen. Excessive use of ink should be avoided and particular care should be taken to see that none gets on the pen arm, otherwise it will become corroded and brittle. The bearings should occasionally be lubricated with a touch of clock oil. This, together with occasional attention to the gate suspension to ensure that the pen always rests very lightly on the paper, should suffice to avoid the development of excessive friction.

Any metal pen should be washed in methylated spirit whenever it shows signs of needing it. The point should be fine enough to produce a thin trace but not so fine as to scratch or stick in the paper. A new pen, if excessively sharp, may be improved by drawing the point once or twice along the side of a safety matchbox.

In the Meteorological Office the preferred type of pen consists of an ink reservoir fitted with a fibre nib. This has the advantage of not requiring constant attention and avoids the problems, already mentioned, arising from too much ink and possible corrosion of the pen arm. Each pen lasts about a year or more.

It is important to remember that, whichever type of pen is used, the distance between the point and the arm gate suspension is the same as printed on the barograph chart in use. In the case of a metal pen, care should be taken to ensure that it is the correct size for the arm.

7.7.7. Evaluation of pressure from the record. In the course of any investigation into some event that has occurred it may be necessary to evaluate the pressure at some particular time by reference to the barogram; this usually arises because the event may have occurred between actual readings of the aneroid barometer. To do this it is necessary to:

(*a*) ascertain the clock error at the particular time by examining the time marks and interpolating if the error was not constant;

(*b*) read the pressure from the barogram as shown by the scale on the chart at the required time, allowance being made for any clock error; and

(*c*) correct for any error in the setting of the pen or the scale value of the chart, or both.

If the correct Meteorological Office chart is used on the instrument there should be no appreciable error in the scale value. In this case, item (*c*) consists only of applying the correction for error in the setting of the pen, which is readily ascertained by comparing the chart readings with the barometer readings at the fixed hours of observation. If the error so obtained is reasonably constant both for high and low readings there is no scale error. If it is not constant there is a scale error as well as a pen-setting or index error. It is then necessary to estimate the actual correction applicable to the recorded pressure at the appropriate time and apply this to the reading shown on the chart.

CHAPTER 8

TEMPERATURE AND HUMIDITY

8.1. TEMPERATURE SCALES

The Celsius scale of temperature has been adopted by the World Meteorological Organization and it is now used by the Meteorological Office. It is replacing the Fahrenheit scale for everyday use in the United Kingdom.

Celsius temperature, t, is defined in terms of the Kelvin scale by:

$$t = T - T_0$$

where T is the temperature in kelvins (K) and $T_0 = 273 \cdot 15$ K. The kelvin is the International System (SI) unit of thermodynamic temperature and is defined as the fraction $1/273 \cdot 16$ of the temperature of the triple point of water.

The practical realization of any temperature scale depends on the acceptance of certain universally defined 'fixed points'. For the Celsius scale it has been internationally agreed that the triple point of water (0·01 °C) and the boiling point of water at a standard pressure of 1013·25 mb (100 °C) shall be two such 'fixed points'. On such a scale the freezing point of water is 0 °C.

8.2. THERMOMETRY

8.2.1. Thermometers. The most familiar form of thermometer is the ordinary sheathed pattern in which the temperature is indicated by the position of the end of the liquid column with reference to a scale of degrees engraved on the stem (see Figure 9A). In the United Kingdom mercury is generally used, except for minimum thermometers in which ethanol (ethyl alcohol) is used.

8.2.1.1. *Accuracy.* All thermometers used by the Meteorological Office for measuring dry-bulb and wet-bulb temperatures were at one time certified by the National Physical Laboratory (NPL) and the Meteorological Office certified the accuracy of all other thermometers. A certificate was issued by NPL giving the corrections to be applied at different points on the temperature scale. The Meteorological Office now exercises strict quality-control during the acceptance of thermometers which renders the application of individual scale corrections unnecessary for many meteorological purposes: for example, the measurement of air temperatures with ordinary sheathed thermometers in naturally ventilated screens, the measurement of maximum and minimum temperatures, and ground surface and soil temperatures. Scale corrections are still seen as necessary for inspectors' thermometers, for the thermometers used in aspirated psychrometers, and when more than usual accuracy is required; in these cases a certificate showing the scale correction is issued with each thermometer, as appropriate.

Ordinary sheathed thermometers, for which no Meteorological Office certificates are issued, are divided into two categories, A and B, by the Meteorological Office Test Laboratory. For easy identification, category A

110

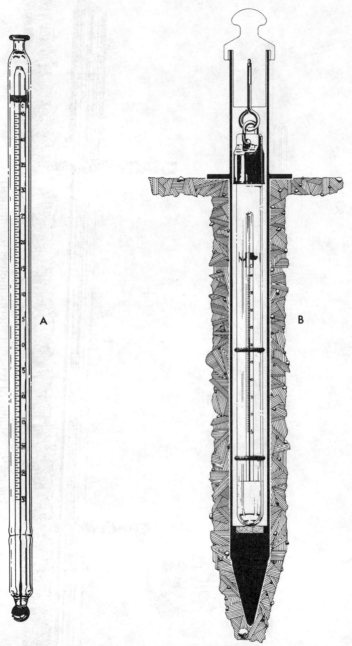

Figure 9.   Standard types of thermometers

A   Sheathed-pattern dry-bulb or wet-bulb thermometer
B   30 cm soil thermometer and steel tube, usually
        installed under a grass-covered surface

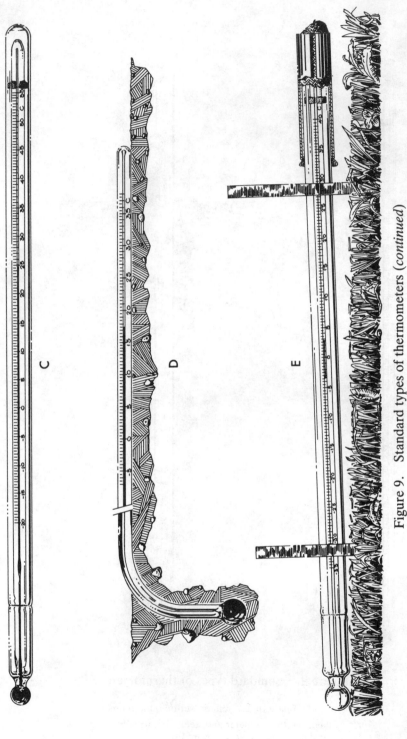

Figure 9.   Standard types of thermometers (*continued*)

C   Sheathed-pattern maximum thermometer   D   Soil thermometer, installed in bare soil   E   Sheathed-pattern minimum thermometer
(illustrated as exposed over grass)

thermometers will be marked with a red paint on the top button and category B will be left unmarked. Any pair of these thermometers in the same category may be used for measuring dry-bulb and wet-bulb temperatures. Category A thermometers will normally be used at all Meteorological Office stations, all climatological stations and all auxiliary stations which render climatological returns and/or full synoptic reports. Category B thermometers will be used at all other auxiliary stations.

As part of the station inspection, the visiting inspector will check all thermometers, both those in use and those held as spares. Thermometers are compared with each other and with an inspector's thermometer in a bucket of water; the water should be at or near ambient temperature but not more than 20 °C and should be well stirred. Any thermometers whose readings differ from the inspector's thermometer by more than the specified tolerances will be reported to the appropriate authority for the necessary action to be taken; normally they will be withdrawn from use.

Maximum and minimum thermometers are difficult to check. Their readings should be noted before they are removed from the screen and the maximum thermometer should be reset. They may then be compared with an inspector's thermometer in a bucket of water (slightly warmer than the temperature to which the maximum thermometer was reset), as described in the preceding paragraph. The thermometers should be dried and reset to their original readings on being returned to the screen. As these types of thermometer are designed to be used in a near-horizontal position, this method of testing does not always give conclusive results. The observer should therefore check that the maximum and minimum thermometers agree with the dry-bulb thermometer when the former have been reset in the shade and replaced in the screen.

8.2.1.2. *Application of corrections.*   As indicated in 8.2.1.1, certificates are issued for certain thermometers, giving statements of the correction to be applied at different points on the scale.

A plus correction means that the stated correction is to be added to the observed reading; a minus correction means that the stated correction is to be subtracted from the observed reading. The errors arise mainly from variations in the bore of the tube, and it may be assumed that there are no sudden discontinuities. In order, therefore, to convert the given corrections (which are only given at a few well-spaced points) into a correction table for everyday use, the observer should first make a rough graph of the corrections given on the certificate and then read off the ranges over which the corrections apply.

Example: the certificate for thermometer No. Met.O. 12345/78 gives the following corrections:

at -30.0 °C	-0.2 °C	at 10·0 °C	0·0 °C
at -20.0 °C	-0.1 °C	at 20·0 °C	$+0.1$ °C
at -10.0 °C	0·0 °C	at 30·0 °C	0·0 °C
at 0·0 °C	0·0 °C		

These values were graphed and the following table was prepared:

-25.1 °C or below	subtract 0·2 °C
-25.0 °C to -15.1 °C	subtract 0·1 °C
-15.0 °C to 15·0 °C	no correction
15·1 °C to 25·0 °C	add 0·1 °C
25·1 °C to 30·0 °C	no correction.

On those occasions requiring the additional accuracy of a certified thermo-meter, the corrections should be applied before the readings are entered in Registers or other form of record.

8.2.1.3. *Precautions in reading thermometers.* The thermometers in the screen are to be read carefully, but quickly, so that the screen door is open for the shortest possible time and changes in temperature due to the presence of the observer are kept to a minimum. To avoid errors of parallax, the observer must read the thermometer in such a way that the straight line from his eye to the end of the liquid column (or index in the case of a minimum thermometer) is at right angles to the thermometer stem. When reading a thermometer the observer must first estimate the tenths of a degree, then note the whole degrees, and enter the temperature in the appropriate space on the observa-tion pad. After all the necessary temperatures in the screen have been noted, a second look should be taken at the thermometers in turn in order to check that the tens and unit figures are correct.

It the thermometers are read at night a low-wattage lamp in waterproof housing may be provided well clear of the screen, perhaps fixed to the enclo-sure fencing. Alternatively a battery-operated torch may be used. Mains lamps must not be used inside the screen, neither must matches nor cigarette lighters be used as a source of illumination.

8.2.1.4. *Estimation to tenths of a degree.* The simplest case occurs when the end of the index or the column of liquid in the bore of the tube coincides exactly with a degree division such as 17, in which case the reading is recorded at 17·0. The decimal point and zero should always be inserted, to make it clear that the entry is not a 'rounded' value, and that the reading was in fact estimated in tenths of a degree. The next easiest case is that in which the reading is just half-way between two degree divisions giving the tenths figure as 5; thermometers calibrated in degrees Celsius have each half-degree divi-sion marked. The estimation of the intermediate tenths is a little more dif-ficult, but four more values can be obtained from the following rules:

just over the exact degree   = ·1
just under the half degree   = ·4
just over the half degree   = ·6
just under the exact degree   = ·9

This leaves only ·2 and ·3, which are respectively a little under and a little over one-quarter, and ·7 and ·8 which are respectively a little under and a little over three-quarters. As a further aid, Figure 10 shows the exact appear-ance of each tenth of a degree from ·1 to ·9.

When the temperature falls below zero, the numerical value of the negative temperature increases towards the bulb end of the thermometer. With nega-tive temperatures the diagram of tenths (Figure 10) needs to be figured in reverse: if the top graduation represents 0 °C, the readings shown in the diagram are (from left to right) −1·0, −·9, −·8 . . . −·1.

Nearly all observers have a personal bias in favour of certain particular numbers. One observer, for example, will hardly ever record ·1 or ·9, prefer-ring ·0, ·2 or ·8; another will nearly always record ·7 in preference to ·6, or ·3 in preference to ·2. Each observer should be on his guard against such personal errors and should try to detect and correct them.

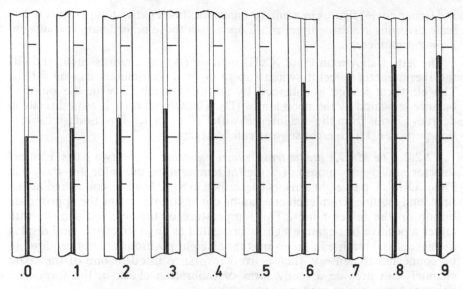

.0   .1   .2   .3   .4   .5   .6   .7   .8   .9

Figure 10.   Estimating tenths of a degree

8.2.2. Electrical resistance thermometers. The principal advantage of electrical resistance thermometers is that they can be connected to an indicator or
recorder up to 1 km away, thus enabling temperatures to be displayed from a
remote site whose accessibility is limited or where the observer's other duties
limit the frequency with which the screen may be visited. The thermometers
contain a small coil of platinum wire whose resistance varies with temperature. In practice, over the relatively narrow band of surface meteorological
temperatures, the relationship is approximately linear and no corrections for
non-linearity are made. The resistance element is placed in a stainless-steel
sheath, near to the tip, to form a robust thermometer with sensing characteristics similar to those of the ordinary thermometers.

Electrical resistance thermometers should be installed at the same height
and location as the ordinary thermometers, care being taken to ensure that
the steel sheath is not in accidental contact with the screen frame. Where both
ordinary and resistance thermometers are in use in the same screen the resistance themometers are mounted behind the ordinary thermometers. A wet
bulb is obtained by using a suitable tubular cotton wick. Procedures for use
and care of electrical resistance thermometers to obtain wet- and dry-bulb
temperatures are closely similar to those for ordinary thermometers and sections 8.4.1, 8.4.1.1. and 8.4.1.2 will provide adequate guidance.

8.2.2.1. *The Mk 5 temperature indicator.* In this temperature indicator
(Plate XXIII) the thermometer inputs are switched by means of a rotary
switch to a low-frequency a.c. bridge circuit which is balanced manually by
the operator. The balance point is found by using a potentiometer linked to
the dial and indicated by the simultaneous glowing of two neon indicator
lamps. The dial has two pointers, in a similar fashion to a clock, with the
smaller one pointing to tens of degrees and the larger to whole degrees. The
dial is scaled with black (outer circle) figures for positive temperatures and

red (inner circle) figures in a mirror image for the negative temperatures. The scale divisions are three times more open than those on ordinary mercury-in-glass thermometers.

The letters shown on Plate XXIII indicate (A) the on/off switch, and (B) the thermometer selector switch (up to eight thermometers can be fitted). The electrical bridge is balanced by adjusting knob C, with the point of balance indicated by the neon lamps (D). The temperature is read directly in degrees Celsius from the dial integral with C. The switch E is used for calibration purposes. Up to eight sensors can be accepted.

8.2.2.2. *The Mk 1A digital temperature indicator.* The two parts, channel selector and display units, of a digital temperature indicator are shown in Plate XXIV. Up to eight sensors (described in 8.2.2) can be connected at any time, and readings from each one can be obtained by pressing the appropriate button on the selector unit. The temperature of the selected sensor, with either a positive or negative sign, is presented as an electronic digital display in degrees and tenths. In the event that all eight positions are not required to be connected to sensors, then a fixed resistance fitted to one of the spare channels can provide a ready form of calibration check in the form of a constant temperature reading.

8.2.2.3. *Routine thermometer checking.* When electrical resistance thermometers are used it is advisable to take a daily check reading against the mercury thermometers, preferably at a time when the temperature is changing slowly. If possible, one observer should read the electrical thermometers at a predetermined time just before another observer opens the screen to read the mercury-in-glass thermometers. If only one observer is available, he should read the glass thermometers as quickly as possible after reading the electrical thermometers. The equipment should be considered suspect if the mean difference between the two dry bulbs or the two wet bulbs exceeds 0·15 °C. If any individual dry-bulb readings differ by more than 0·5 °C, or wet-bulb readings by 0·8 °C, then more frequent comparisons should be made to determine if the mean differences have moved outside the limits. If the wet-bulb readings do not agree, and wick lengths are correct, the wicks should be removed and the thermometers rechecked as dry bulbs.

8.2.2.4. *Field calibration of electrical resistance thermometers.* The Meteorological Office issues detailed instructions concerning the calibration of electrical resistance thermometers when first installed and on routine inspections.

8.2.3. Rounding temperatures to the nearest whole degree. When any instructions specify that a temperature reading is to be reported or recorded only to the nearest whole degree the procedure is as follows:

(*a*) Read the thermometer by estimation to the nearest tenth.

(*b*) Add or subtract any necessary correction (see 8.2.1.2).

(*c*) If the number of tenths in the corrected value is 4 or less, disregard the tenths figure; for example:

corrected values: 12·0, 5·1, 15·3, 26·4,
rounded values: 12,   5,   15,   26.

(*d*) If the number of tenths in the corrected value is 6, 7, 8 or 9, add 1 to the number of whole degrees; for example:

corrected values: 2·6, 5·7, 11·8, 18·9,
rounded vlaues: 3, 6, 12, 19.

(*e*) If the number of tenths in a corrected value is 5, 'throw to the odd', i.e. round off to the nearest odd whole number; for example:

corrected values: 6·5, 7·5, 8·5,
rounded values: 7, 7, 9.

The above conventions also apply when rounding negative temperatures, and to the readings from electrical resistance thermometers although, in this case, no estimation of the value of the tenths of a degree is required since this is either displayed or read directly and corrections to the readings are not required.

8.3. THERMOMETER SCREENS

Every effort must be made to ensure that all thermometers used for meteorological purposes are exposed in as near standard conditions as possible. Only when this has been achieved can any real comparison between temperatures be made. This extends to readings made either at the same place or at widely differing geographical locations.

8.3.1. Approved screens. The thermometers used to determine the air temperature, the humidity and the maximum and minimum air temperatures are exposed in an approved type of screen. These screens are designed to shield the thermometers from precipitation and radiation while, at the same time, allowing the free passage of air. This is achieved by constructing the screens with louvered sides and door, a double roof with an air space between the inner and outer components, and a floor consisting of three partially overlapping boards separated by an air space. The screen is mounted on a stand so that the bulbs of the dry- and wet-bulb thermometers (which are mounted vertically) are 1·25 m above ground which ideally should be covered by short grass. As a further measure of standardization all such screens are painted in white gloss so that as much radiation is reflected as possible. This paintwork should be maintained in good condition. Daily familiarity with a screen makes it comparatively easy to accept a slow deterioration in the condition of the paintwork but, if allowed to continue unchecked, such deterioration can eventually lead to doubts as to the validity of any temperature readings made in the screen, and rot in the woodwork.

There are two sizes of thermometer screen. The large one (Plate XXV) can hold two ordinary thermometers and maximum and minimum thermometers (also, where necessary, electrical resistance thermometers) and a bimetallic thermograph and a hair hygrograph (see Plate XXIV). The ordinary screen will not hold any autographic instruments.

Where siting difficulties do not permit the installation of either of these screens it may be possible to use a marine screen fixed so as to have an unimpeded circulation of air through it. Stands are not available for this type of screen. Dry- and wet-bulb mercury thermometers and electrical resistance thermometers only can be used. The screen's notice 'Hang to windward' is a reminder when the screen is used aboard ship.

At mountain stations subject to heavy snowfall an arctic screen may be necessary.

8.3.2. Care of the screen.   Thermometer screens should be brushed out frequently, special care being devoted to the spaces between the louvers where dirt is apt to accumulate. About once a month, or more often in industrial localities or at coastal stations after onshore winds, the screen should be given a good wash with soap and water, or detergent and water.

Cleaning operations are best carried out immediately after an observation hour (preferably 0900 GMT) so that the mercury thermometers, which should be dismounted and put in a safe place while cleaning is in progress, may have time to recover from the disturbance before the next observation is due. The readings of the maximum and minimum thermometers should be noted before dismounting.

Screens must be repainted if washing does not restore a clean white surface, or if the paint has started to peel or crack. There is no fixed period but normally the screen will require repainting about every two years. Rub down well with glass-paper and then apply an undercoating followed by a finishing coat of white hard-gloss paint of exterior quality. Any necessary repairs to the woodwork should be attended to before painting. In intermediate years a single coat of white hard-gloss paint may be applied after light sanding. The instruments should be removed whilst the inside of the screen is painted, but thermometers should be replaced as soon as possible to avoid the loss of any readings. Where it is necessary to leave electrical resistance thermometers in the screen, paint must be kept clear of the sensors.

As the metal stands supplied and recommended by the Meteorological Office are heavily galvanized they need no painting for many years. If the screen is mounted on a wooden stand, the stand should be painted at the same time as the screen.

The screws holding the screen to the stand should be 1¾ inches long to ensure that the screen is secure. In severe gales some screens have been blown off their stands through the use of screws of inadequate length.

8.4. STANDARD THERMOMETERS IN THE SCREEN

Four thermometers are normally exposed in an approved screen: two similar ordinary thermometers and a maximum and minimum. All these thermometers are discussed in the following sections and their arrangement in the screen is shown in Plate XXIV (in which two electrical resistance thermometers are also included). Other thermometers are exposed outside the screen to give further specialist readings (see 8.5).

8.4.1. Dry-bulb and wet-bulb thermometers.   It was recommended earlier (8.2.1.1) that these thermometers should be selected from among those marked with red paint for nearly all stations. The dry-bulb thermometer is exposed 'as it stands' and indicates the temperature of the air at the time of observation. The wet bulb is kept constantly moist by a cotton covering, the end of which dips into a reservoir of purified water. The wet-bulb thermometer indicates the 'temperature of evaporation' which is, in normal circumstances, lower than the air temperature (see 8.4.1.2). The difference between the dry-bulb and wet-bulb temperatures is known as the 'wet-bulb depression'. From the dry- and wet-bulb readings, relative humidity and vapour pressure can be obtained by reference to *Hygrometric tables* (Part II) or by the use of a humidity slide-rule. The combination of dry- and wet-bulb thermometers in a screen is known as a non-aspirated psychrometer (or non-aspirated hygrometer). For synoptic reports dew-point can be obtained from the slide-rule or from the booklet *Dew-point tables for screen readings, degrees Celsius*.

8.4.1.1. *Care of the wet-bulb thermometer*.   In the calculation of dew-point the important element is the wet-bulb depression. It is necessary for the wet bulb to function correctly if this depression is to be accurately determined. The observer must give attention to the following points.

(a) *Use of purified water*.   The water used for maintaining a wet bulb should be purified water. Water obtained when defrosting a refrigerator may be used. Clean rain-water is also permissible in an emergency, but tap-water and sea-water must not be used. Purified water obtained from a garage is liable to be contaminated with sulphuric acid when taken from a bulk source. 'Hard' tap-water will cause a deposit to build up over the bulb of the thermometer. This will cause false readings and any attempt to remove the deposit may lead to a broken thermometer.

(b) *Covering of wet bulb*.   When either ordinary thermometers or electrical resistance thermometers are used as wet bulbs, a tubular cotton wick is used. The wick must either, in the case of an ordinary thermometer, cover the bulb and extend up the stem, or cover the resistance thermometer sensor, and in both cases for a specified length. The length of the wick actually covering the thermometers is important, not only to accord with the physical factors affecting the derivation of humidity, but also to obtain readings which are consistent between each of these two types of thermometer. In both cases the wick should be tied just below either the bulb or the tip of the sensor and again, for an ordinary thermometer, immediately above the bulb. The wick is then continued up the stem for 50 mm for the ordinary thermometer and tied again, 68 mm for Mk 2 electrical resistance thermometers (as shown for both in Figure 11) and 42 mm for Mk 4A electrical resistance thermometers.

Unbleached tubular wick should be given two short boils of 15 minutes in water and a little liquid detergent, with a tap-water rinse between, and then rinsed thoroughly in purified water before use. It may be convenient to keep one length of boiled wick in a sealed container of purified water prior to use, otherwise the wick should be stored dry in a clean plastic bag. The wick must be handled as little as possible, and then only with clean grease-free fingers. Tubular wick issued by Meteorological Office stores is usually bleached before dispatch.

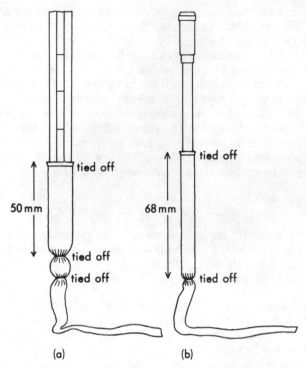

Figure 11.   Wet-bulb coverings for (a) ordinary and (b) electrical resistance
thermometers Mk 2

(c) *Changing the wick*.   The wick should normally be changed once a
week, but it should also be changed immediately after a period during
which even slight dust has been raised. In addition, at coastal stations,
the possibility of the wet bulb being affected by salt spray must be
remembered, and immediately after a storm with onshore winds both
water and wick must always be changed.

   The changing should be carried out at least 15 minutes before a
scheduled observation is due. The fact that the wick has been changed
should be recorded in the remarks column of the Register.

(d) *Water container*.   The water container should have a narrow neck or
be fitted with a perforated cap to allow the passage of the wick; it
should be placed near the wet bulb so that the distance which the water
has to travel along the wick exposed to the air is not excessive. There
should be no slack or dip in the wick, otherwise water may tend to drip
from the lowest point and empty the container. The container should
be replenished daily, or at shorter intervals in hot dry climates, to avoid
any risk of its running dry. Non-breakable plastic containers held to the
base of the screen are preferred.

(e) *Wet-bulb temperature below freezing*.   When the wet-bulb temperature is below freezing-point it is necessary that the wet bulb should have
a thin coating of ice instead of a coating of water. All derivations of

humidity are based on the assumption that the wet bulb is frozen once the temperature of it has fallen below the freezing-point. Ice evaporates in the same manner as water and once an ice bulb has been established it will need to be renewed at intervals. The maintenance of a properly functioning ice bulb is no easy matter. The application of too much water will result in an excessive build-up of ice which will invalidate any readings. If this occurs the excess ice must be cleared by applying lukewarm water, melting the ice and re-establishing a thin ice bulb in sufficient time before an observation for a true ice bulb to be formed. In very dry conditions an ice bulb will need renewing for every observation. Even when a covering appears frozen, a careful application of ice-cold water with a camel-hair brush to moisten the wick completely will be necessary in order to ensure a thin uniform covering of ice over the whole bulb. If the water applied to the wick does not freeze readily, induce it to do so by touching it with a little hoar frost, snow or fragments of ice collected on the end of the camel-hair brush. The same procedure should be adopted on those occasions when the dry-bulb temperature is below freezing but (owing to supercooling) the wet bulb is still liquid. Preparation of the ice bulb may have to be started up to 30 minutes before the observation is due.

When, in exceptional circumstances, it is not possible to induce an ice bulb, the supercooled wet-bulb reading should not be used in computing relative humidity. At those stations with hair hygrographs, the relative humidity should be taken from the record at the time of observation. An appropriate correction should be applied to the hygrograph reading, the correction being based on recent comparisons between the hair-hygrograph record and the corresponding value of humidity computed from dry- and wet-bulb temperatures when the wet-bulb reading exceeded 0·0 °C. Where no hygrograph is available, observers should note the reading of the unfrozen wet bulb without attempting any derivation of humidity.

8.4.1.2. *Exceptional ice- or wet-bulb readings.* Correct functioning of the wet-bulb (or ice-bulb) thermometer is normally indicated by the reading being lower than that of the dry bulb, though the two readings will be the same when the air is completely saturated and the temperature is above freezing-point. With air temperatures below freezing-point it is physically possible, when no water fog is present, for the ice bulb to read up to about 0·3 °C higher than the dry bulb. Such occasions are of great interest, and the observer should check that the ice bulb is functioning correctly before recording the observations as genuine, and make an appropriate note in the remarks column of the Register.

An authentic reading exceeding the dry-bulb reading can occur only with an ice bulb, never with a wet bulb in the ordinary sense of the term. If the wet-bulb temperature is above 0 °C and is higher than the dry-bulb temperature, either the wick on the wet bulb is very dirty and should be changed immediately, or one or other of the thermometers is defective.

Very occasionally, during rapid fluctuations of temperature, the wet bulb may read fractionally higher than the dry bulb owing to the difference in the speed or response. On such occasions, if there is no doubt that the wick is clean and that the thermometers are not defective, the wet bulb should be taken as correct and the dry-bulb reading adjusted to conform with the wet bulb.

8.4.2. Maximum thermometer.   This thermometer (Figure 9C) depends for its action on the presence of a small restriction in the bore of the tube below the lowest graduation near the bulb end. The constriction offers considerable resistance to the flow of mercury. The pressure created by the expansion of the mercury with rising temperature is sufficient, however, to force mercury past the constriction. When the temperature falls, provided that the stem of the thermometer is approximately horizontal, no reverse force comes into play, the mercury column breaks at the constriction, and the mercury in the tube beyond the constriction is left behind. If after such a fall in temperature there is a rise to a temperature higher than that existing before the fall, additional mercury will be forced past the constriction. Thus the reading of the far end of the mercury column in the tube indicates the highest temperature reached since the thermometer was last set.

The thermometer is installed in the thermometer screen upon the metal clips on the thermometer support, inclined at an angle of about 2° to the horizontal with the bulb slightly below the stem. This arrangement ensures that the mercury thread does not run up the stem away from the constriction.

After reading the maximum thermometer at the specified time it should be reset, except at the reading for 6 p.m. Press reports during the period of British Summer Time when the thermometer is not reset. The method of setting involves forcing the mercury back through the constriction until the mercury forms a continuous thread from the bulb to the meniscus. In order to do this, the thermometer should be gripped firmly by the centre of the stem and swung vigorously up and down in a semicircle, taking care that it does not slip out of the hand; (do not hold the thermometer in a gloved hand). Care should also be taken to avoid striking an obstacle, including the observer's own person or clothing.

When set, the maximum thermometer should show a reading agreeing with the dry bulb; any difference between the readings of the two thermometers should not exceed the following figures:

$\pm 0{\cdot}2$ °C when checked with a Category A ordinary thermometer ⎱ with dry-bulb temperature
$\pm 0{\cdot}4$ °C when checked with a Category B ordinary thermometer ⎰ above 0 °C

$\pm 0{\cdot}3$ °C when checked with a Category A ordinary thermometer ⎱ with dry-bulb temperature
$\pm 0{\cdot}5$ °C when checked with a Category B ordinary thermometer ⎰ below 0 °C.

The thermometer is then replaced on its supports. The bulb end is replaced first and the stem is then carefully lowered until the other end rests on its support.

Maximum thermometers occasionally develop a tendency to act as ordinary thermometers; that is to say, with falling temperature, the mercury tends to move back against the resistance of the constriction. This fault cannot be cured by the observer and the instrument should therefore be replaced.

8.4.3. Minimum thermometer.   The minimum thermometer is a sheathed thermometer of the spirit-in-glass type with a small dumb-bell shaped index of dark glass in the bore. This index is prevented from breaking through the meniscus of the spirit column by surface tension. As the temperature falls, the index is drawn towards the bulb, but it remains stationary as the temperature rises. The end of the index furthest from the bulb therefore indicates the minimum temperature reached since the last setting.

The thermometer is installed in the thermometer screen upon the metal clips on the thermometer support, inclined at an angle of about 2° to the horizontal with the bulb slightly below the stem. With this arrangement the movement of the index towards the bulb is slightly assisted by gravity.

The reading of the minimum is taken by noting the position of the extreme end of the index nearest to the meniscus. Setting is carried out by lifting the thermometer off the clips and tilting the thermometer bulb upwards so that the index slides along the tube and makes contact with the meniscus. When replacing the thermometer in the screen, incline the instrument with the bulb upwards and place the other end on its support first.

After the thermometer has been reset and replaced in the screen it should be read again and the reading compared with that of the dry-bulb thermometer. The difference between the readings of the two thermometers should not exceed the values given in 8.4.2.

8.4.3.1. *Errors in minimum thermometers.* If the comparison of temperatures mentioned above shows that the minimum thermometer reads consistently lower than the given limits, the cause is probably due to detached spirit. Unlike mercury, spirit wets the glass and consequently, if the temperature falls rapidly, a thin film of spirit may be left on the walls of the bore, or the same effect may be caused in hot weather by distillation of spirit from the liquid column to the upper walls and even to the safety chamber at the top of the bore. Bubbles may also form in the spirit column during transit.

The spirit column can usually be reunited by bouncing the bulb end of the thermometer on a thick telephone directory or pile of newspapers laid flat on a table. The thermometer should be held vertically, with the bulb end downwards and about 12 cm above the directory, and then let slip through the fingers so that the bulb end bounces on the directory. This process should be continued until all the bubbles have gone and the spirit column shows no breaks. This treatment may not always be sufficient, especially if spirit is present in the safety chamber; in that case the thermometer should be placed vertically, bulb downwards, in a vessel filled with water sufficiently hot to cause the liquid column to expand to the safety chamber, *but not so hot that the chamber is filled with liquid* (a convenient criterion is that the water should be just too hot to keep one's finger in) and the vessel with the thermometer allowed to cool slowly for 24 hours. Whichever method is used it is most important that the thermometer should be left standing upright, bulb end downwards, for 24 hours immediately after the treatment, otherwise bubbles are likely to reappear. A spare rain-gauge bottle or similar narrow-necked bottle makes a good holder.

Bubbles may occasionally develop in the stem of the minimum thermometer in the screen, even though the instrument has not been disturbed in any way and has been inspected and found in order on the previous evening. This is especially so on cold nights when the minimum temperature recorded is well below freezing-point. The correct procedure for such occasions is:

(*a*) Discard the reading.

(*b*) Take the recommended steps to remove the bubbles before resetting.

(*c*) Make a note of the reading, the condition of the thermometer and the action taken, in the Register and the monthly return.

8.4.4. Procedure for reading the thermometers.   At most synoptic stations and all climatological stations, the dry-bulb and wet-bulb thermometers are read at every hour for which an observation is made. At stations which are also required to make special reports for aviation purposes, additional readings of these temperatures are required for the half-hourly reports. In addition, the maximum and minimum thermometers are read at the hours specified in Chapter 1 for different types of station, and are reset immediately after they have been read. See, however, the special instructions in 1.5.2 (page 8) applicable to those stations making special observations at 6 p.m. clock time. Special procedures are arranged at some auxiliary reporting stations.

When all four thermometers have to be read, the order of reading should be dry bulb, wet bulb, maximum, minimum.

8.4.5. Care of thermometers in the screen.   Moisture may be deposited on all surfaces inside and outside the thermometer screen by condensation of water vapour in wet fog or from some other cause. This deposit of moisture can cause significant errors in the readings of thermometers if the bulbs of those thermometers intended to be dry become moist; errors will continue until all the moisture has evaporated (which may not be for several hours after the conditions which led to the moisture has ended). The dry-bulb thermometer would act as a partial wet-bulb thermometer and would indicate a lower temperature than a true dry bulb. Consequently the derived dew-point using this reading in conjunction with the wet-bulb reading would also be in error. Similarly the maximum and minimum thermometers might indicate lower temperatures than the true values if their bulbs are moist at the time that the true maximum or minimum temperatures are reached.

If the dry-bulb, maximum or minimum thermometers are affected in this manner, the deposits of moisture should be removed with a clean tissue, blotting-paper or cloth, care being taken not to warm the thermometer in the process. A note of the time and action should be made in the remarks column of the Register.

8.4.7. Discrepancies in readings.   If all four thermometers in the screen are working correctly it is clear that:

(*a*) The reading of the maximum should be higher than (or at least as high as) any reading of the dry bulb since the maximum was last set.

(*b*) The reading of the minimum should be lower than (or at least as low as) any reading of the dry bulb since the minimum was last set.

If the readings of the maximum and minimum have been checked at the time of observation and are then found to be inconsistent with the above rules, the cause of the discrepancy may be either an error in reading the dry bulb at one of the hours during the period covered, or a defect in one of the thermometers. The reading which, on examination, the observer considers to be erroneous should be adjusted by him, and the adjusted value should be entered in the Register within brackets. A note of the actual reading should also be made in the remarks column, both in the Register and in the appropriate monthly return.

In winter the night minimum temperature often occurs around dawn. In these circumstances, especially at synoptic stations where the first reading of the minimum is taken at 0600 GMT, if the fall of temperature is rapid and is still in progress, the minimum thermometer may show a slightly higher reading than the dry bulb simply because the lag of the minimum thermometer is greater than that of the dry bulb. In this instance it is legitimate to adjust the minimum reading to agree with the dry-bulb reading.

At observing stations on or very near to aerodromes, errors in thermometer readings may be caused by the hot exhaust gases from jet aircraft, either standing with engines running near the meteorological enclosure or even taxiing past it. The gases may be projected towards the enclosure by the jets, or they may be carried to it by the wind after they have lost their original momentum.

With dry-bulb and wet-bulb thermometers the only precaution necessary is to avoid reading them during, or immediately after, exposure to exhaust gases. With a maximum thermometer, however, it is necessary to consider whether it may have been affected in this way at any time since it was last set. At observing stations where a thermograph is in use the question can usually be decided by an examination of the thermograph; sharp kicks in the trace, caused by the passage of a taxiing aircraft or by the running of an engine nearby, can be easily recognized. If there is no thermograph, the maximum temperature should be checked by comparison with the half-hourly or hourly temperatures, if available. There should be fairly close agreement between the highest of these and the readings of the maximum thermometer.

Any maximum thermometer reading which the observer considers erroneous should be adjusted and entered in the Register, as indicated above for other suspected temperature readings.

8.5. THERMOMETERS EXPOSED OUTSIDE SCREENS

Additional thermometers are required to be exposed outside the screen, usually on or near to the surface of the ground and, in special circumstances, below the ground (soil thermometers) at specified depths. Readings from these thermometers are required for particular purposes and care must be taken to ensure that, where necessary, they are set correctly and that no confusion arises when entering the recorded temperatures in the appropriate Register or return form.

8.5.1. Grass minimum thermometer. The grass minimum thermometer (see Figure 9E) is similar to the minimum thermometer used in the screen. It is used to record the lowest temperature reached during the night over short grass freely exposed to the sky. It is read at 0900 GMT daily, but full synoptic stations also read the thermometer at 0600 GMT without resetting.

The most open position available within the enclosure should be selected for the grass minimum, and the grass should be kept short so that its condition resembles that of a lawn. The thermometer should be laid on two Y-shaped pegs or placed in the special grass minimum thermometer supports which are made of stiff black rubber. When the pegs are used they should support the thermometer at an inclination of at least 2° to the horizontal so that the bulb is lower than the stem and at a height of between 2·5 and 5 cm above the ground, with the bulb in contact with the tips of the grass. If the special rubber supports are used, the thermometer is pushed through the hole in the rubber squares, the large one adjusted to be about 29 cm from the bulb and the smaller square 11 cm from the bulb, thus ensuring that the thermometer will be at the required inclination to the horizontal. By choosing a particular side of the squares to rest on the ground, the thermometer bulb will be just touching the tips of the grass.

The thermometer is set by tilting the bulb above the stem so that the index runs down the bore until it comes into contact with the meniscus. The thermometer should be set and exposed at the last observation hour before sunset. For example, if sunset is at 1935 GMT, then a station making hourly observations throughout the 24 hours will set and expose the thermometer at 1900 GMT; a station observing only at main and intermediate synoptic hours will set and expose at 1800 GMT, and a station closing down at 1700 GMT will set and expose at that time. At stations where observations are only made at 0900 GMT and the thermometer is reset and exposed more than 2 hours before sunset, it is essential that an anti-condensation shield be fitted.

An anti-condensation shield is a black metal sheath fitted over the thermometer at the end remote from the bulb. The black sheath absorbs more heat from the sun than the exposed part of the thermometer so vapour is prevented from condensing in the upper part of the bore, and any vapour which has already condensed there evaporates and condenses lower down the stem at the top of the ethanol column. The shield should be repainted with black paint when necessary.

Except at the stations where the grass minimum is reset and left exposed during the day, the thermometer should be stored after the 0900 GMT reading, the most convenient place being the thermometer screen. A thermometer taken off wooden pegs should be stored in a near-vertical or sloping position with the bulb end lower. A thermometer with rubber supports attached may be stored in the well in the floor of the screen. If for any reason the thermometer cannot be stored in the screen, it should be brought indoors and stored in a position as near vertical as is practicable, with the bulb end downwards.

The grass minimum thermometer should be checked with the screen dry-bulb thermometer about every 3 months. If the grass minimum thermometer is stored in the screen on the special rubber supports (i.e. at the correct angle), the check may be carried out during the day.

As with all ethanol-type thermometers, bubbles or breaks in the liquid column are liable to occur, usually under conditions of great cold or when the thermometers are left out in the sunshine. The effects of this latter cause can be alleviated by the use of the anti-condensation shield described above. If such bubbles are seen when the thermometer is due to be read, no reading should be taken.

The methods for removing bubbles from minimum thermometers are described in 8.4.3.1 (page 123).

8.5.1.1. *Procedure with snow on the ground.* When snow covers the ground the thermometer should be supported immediately above the surface of the snow, as near to it as possible but not actually touching it, using the wooden pegs or the rubber supports. If, owing to a fall of snow overnight, the thermometer is found to be buried in the morning at the time of reading, the snow should be carefully removed and the thermometer read in the usual way. The reading should be entered in the Register and marked '?', and a note 'grass minimum buried in snow' entered in the remarks column. If the thermometer is to be left exposed it should be replaced at the snow surface as described above. Doubtful readings should not be included in synoptic messages.

At synoptic stations where the enclosure is visited during the night hours the thermometer should not be left buried in the snow until it is first due to be read. When the snowfall appears to have ceased the thermometer should be supported above the snow surface, care being taken not to displace the index, so that a reading of the temperature immediately above the snow surface may be available at the observation time. The time when the thermometer was placed above the snow should be noted in the remarks column of the Register. If, having carried out the above procedure, further falls of snow occur at intervals and may be doing so at the time of observation, the snow should be cleared and the temperature read from the undisturbed thermometer. A note on the extent to which the thermometer was covered should be made.

8.5.2. Ground minimum thermometers. The same type of thermometer as the grass minimum (described in 8.5.1) is also used at selected stations for measuring minimum temperatures on bare soil and/or on a concrete slab. Suitable sites for bare-soil and concrete-slab minimum thermometers are indicated in Figure 20 (page 179), and details of exposure are given in Appendix I in I.8.7 and I.8.8 respectively. The same sites, once selected, must always be used.

The time for setting and exposing both these thermometers is the same as for the grass minimum, i.e. the last observation hour before sunset, as detailed in 8.5.1. Each thermometer should be read at 0900 GMT daily, and they are then stored in the screen or indoors as advised in 1.10.3. When left exposed during the day or exposed more than two hours before sunset the thermometers must be fitted with an anti-condensation shield.

Concrete-slab and bare-soil minimum thermometers should be checked against the screen dry-bulb temperature at approximately 3-monthly intervals, using the same tolerances as given in 8.2.1.1. The possibility of this type of thermometer developing bubbles should be borne in mind and action taken, as described in 8.4.3.1, if it occurs.

8.5.2.1. *Bare-soil minimum thermometer.* The thermometer should be set, then laid on level bare soil with the stem having a slight slope downwards towards the bulb. Small pegs of wood should be placed in the ground at each side of the thermometer, but not near to the bulb, to prevent accidental movement.

When snow covers the ground the thermometer should rest on top of level snow. When the thermometer is buried by overnight snow it should be carefully extricated and then exposed on top of level undisturbed snow as close as possible to the normal site. Appropriate notes should be made in the remarks column of the Register, and the procedures of 8.5.1.1 followed.

8.5.2.2. *Concrete-slab minimum thermometer.* The thermometer should be set and then placed in the PVC-coated phosphor-bronze spring clip with the end of the anti-condensation shield just touching the clip and with the thermometer stem parallel to the major axis of the slab. The thermometer bulb will then be in the centre of, and in contact with, the slab and the thermometer will have the correct downward slope of approximately 2° (see I.8.8 in Appendix I).

When snow covers the ground, the concrete slab should be swept clear of snow at the time of setting the thermometer. If there is a fall of snow overnight this should be removed carefully as soon as possible after the snow has ceased, and the free exposure of the thermometer restored as far as possible, care being taken not to displace the thermometer index. If the thermometer is, nevertheless, covered by snow at 0900 GMT, the snow should be removed carefully and the thermometer read in the usual way. Appropriate notes concerning the removal of snow from the concrete slab or about the thermometer being covered by snow at 0900 GMT should be entered in the remarks column of the Register and the procedures of 8.5.1.1 followed.

8.5.3. Soil thermometers.
At certain selected stations the temperatures below the level of the earth's surface are measured at various depths. The depths used in the United Kingdom are 5, 10, 20, 30 and 100 cm, although measurements at all of these are not necessarily taken and, exceptionally, temperatures at other depths may be measured. To cope with these variations in depths, two special types of thermometer are used: one type for depths of 30 cm or more and the other for depths of less than 30 cm.

8.5.3.1. *Soil thermometers for depths of 30 cm or more.* The type of thermometer used for these depths is shown in Figure 9B. The thermometers are enclosed in glass tubes and their bulbs are embedded in wax to make them insensitive to sudden changes of temperature. This allows them to be drawn to the surface and read before their temperature has had time to change appreciably. As underground changes of temperature are very slow, the slow response resulting from the coating of wax does not lead to inaccuracies in the measurements.

At depths of 30 cm or more, the temperatures are measured under a grass surface. The thermometers are suspended in steel tubes sunk through the surface of the grass plot.

Water must not be allowed to collect in the steel tubes; to prevent this, the tubes are fitted with caps. The thermometer is suspended from the cap. If water is found to be present, it can be removed by means of a sponge or other absorbent material tied to the end of a stick. Leaking tubes should be repaired and faulty caps replaced.

Temperatures should be read to the nearest tenth of a degree; to avoid parallax errors the thermometer should be raised to eye-level (see 8.2.1.3). The instrument should be shielded from direct sunshine during the observation.

8.5.3.2. *Soil thermometers for depths of less than 30 cm.* These thermometers (pattern for use at 5 cm is shown in Figure 9D) are unmounted and

unsheathed with a bend in the stem between the bulb and the lowest graduation. The bend allows the bulb to be at 5, 10 or 20 cm, according to the
particular pattern, when the vertical part of the stem is sunk into the ground
with the horizontal (graduated) part of the stem in contact with the surface.
Measurements at these depths are made under a bare soil surface. The thermometers are read without being disturbed from their position in the soil, the
readings taken to the nearest tenth of a degree.

Instructions for installation under bare soil are given in I.8.5, page 187, and
these should be carefully followed if it is necessary to move or replace the
thermometers at any time.

In dry weather, cracks in the soil develop, especially in clay, and it sometimes happens that the vertical section of a soil thermometer is situated in a
crack. The development of cracking should be delayed, when warm dry
weather is expected, by light raking of the surface soil to prevent it becoming
smooth and hard. This treatment must be carried out, with as little disturbance as possible, at any time after rain when the soil is partially dry and in a
suitable condition. In this way cracks which have appeared may sometimes be
got rid of, but in a prolonged dry spell nothing should be done; the readings
should be accepted. The presence of the cracks is recorded in the observation
of state of ground (see 6.2.1).

8.6. AUTOGRAPHIC INSTRUMENTS

8.6.1. The bimetallic thermograph (Plate XXVI) gives a continuous record of
air temperature. For routine use the thermograph is accommodated in the
large thermometer screen (Plate XXV), positioned to the left of the thermometer supports on the middle bottom board of the screen.

The temperature-sensitive element is a bimetallic helix formed by welding
together strips of two metals which have widely different coefficients of expansion; changes of temperature cause the helix to coil or uncoil. In the
thermograph this action is made to control the movement of a recording pen,
the pen-arm spindle being attached directly to one end of the spiral, the other
end of which is anchored to the framework of the instrument. The mechanism
is thus very simple, and as the spiral exposes a large surface to the air the
instrument has a lag about half that of the ordinary mercury thermometer and
is capable of recording short-period as well as long-period variations of
temperature. A weekly or daily clock can be fitted to drive the drum on which
the chart is secured.

The procedure for changing the chart and putting time marks on the record
is the same as for the barograph (see 7.7.3 and 7.7.4, pages 107 and 108, and
also 1.1.4, page 4). Time marks should be made gently, displacing the pen by
the equivalent of only one or two degrees of temperature on the chart.

Two chart ranges are in use in the British Isles: the winter chart has a range
from −25 to +20 °C and the summer chart from −10 to +35 °C. A third chart
range of +10 to +55 °C is used at certain overseas stations. The change from
the winter to the summer chart should usually be made at 0900 GMT on the
Monday nearest the middle of April, and the change from summer to the
winter chart at 0900 GMT on the Monday nearest the middle of October, but

the observer should use his own discretion in the matter. Abroad, the best dates for changes will depend on the locality and the range of charts provided.

When changing from summer to winter charts, or vice versa, the position of the pen arm has to be adjusted to register the correct dry-bulb temperature. Coarse adjustment of the position of the pen on the new chart is effected by loosening the screw which attaches the pen arm to the pen-arm spindle, rotating the pen arm about the spindle until the pen records approximately the correct dry-bulb temperature, and then tightening the screw. The pen is brought to the correct dry-bulb temperature by use of the fine-adjustment screw, which is the milled-head screw at the fixed end of the helix.

This final adjustment to the pen is best done on a cloudy day when the temperature is steady. The thermograph should be left in the screen for half an hour and then compared with the reading of the dry bulb. Alternatively, a room with a constant temperature may be used. Place the thermograph away from direct sunlight and draughts, and set a thermometer with its bulb as close as possible to the bimetallic helix. Some instruments have a clip on the guard for holding a thermometer.

8.6.2. The hair hygrograph.   This instrument (Plate XXVI) is used to obtain a continuous record of relative humidity. For routine use, the hygrograph is accommodated in the same large thermometer screen as the thermograph, positioned to the right of the thermometer supports.

The sensitive element is a bundle of human hair from which all oil has been extracted. The length of hair so prepared is affected by changes in the relative humidity of the air, but is practically unaffected by temperature. The length increases with increasing relative humidity but the rate of increase is not uniform throughout the range. For example, a change from 90 to 95 per cent gives a much smaller change in length than a change from 40 to 45 per cent. The hair is clamped between two rigid supports and kept taut by a loaded hook at the middle. The movement of the hook is transmitted through a lever and two quadrants to the pen arm, the quadrants acting as a lever of varying length; in this way the non-linear response of the hair is changed to a linear response on the chart. The pen arm is fitted with a gate suspension and the pen records on a chart wound on a clock-driven drum which may be daily or weekly as required.

The hygrograph should be tapped slightly with a finger before a reading is taken.

The procedure for changing the chart and making time marks is detailed in 7.7.3 and 7.7.4, pages 107 and 108). In making time marks it is important that the pen arm should be moved downwards about 3 mm; an upward movement strains the hair.

8.6.3. Checking, adjusting and maintenance of thermographs and hygrographs.   Certain procedures are common to both types of autographic instruments. A check should be made on the setting of the pen whenever a chart is changed; this is to ensure that the pen starts recording on the chart at either the temperature of the dry bulb in the screen or, in the case of the hygrograph, at the humidity at that time, to values as close as adjustment will allow. A further method of checking the setting of a hygrograph is outlined below in 8.6.3.2. When the charts are finally removed, the values of temperature and humidity obtained from the screen thermometers should be entered against the time marks on the charts and a comparison made between the actual values and those recorded on the chart. A small, near constant, difference between the two can usually be removed by using the fine adjustment. A non-constant difference over a week between actual and recorded values may be due to other causes: dirty hairs on the hygrograph, a distortion of the chart drum, or the drum may not rotate in the horizontal are possible reasons. The last two can be checked by manually rotating the drum once, at either a constant temperature or humidity, with a chart in position and the pen recording. The trace so obtained should be horizontal. Experience has shown that frequently this is not the case and, on thermograph traces, errors of up to 0·9 °C have been detected. If such cases are found, the fact should be noted in the instrument log (see 1.10, page 14) and reported to the appropriate authority for a decision as to whether the instrument should be exchanged. In the particular case of the hygrograph, unless the instrument is well maintained, errors due to drum distortion or the drum not rotating horizontally may be compounded by errors arising from the mechanical malfunctioning of the instrument movement or dirty hairs. Dirt on the clock and drum collars is often the cause of trouble.

The timing of the clocks should be checked by comparison between time marks. Alterations can be made to the clocks by removing them from the instrument and adjusting the regulator in the base. The daily clock allows a total adjustment of 15 minutes and the weekly clock a total adjustment of 1¾ hours.

If metal pens are used, they should be cleaned of sediment as necessary with Inhibisol or methylated spirit. The instructions (for barograph pens) in 7.7.6, page 109, should also be followed. The type of pen recommended for the thermograph consists of an ink reservoir fitted with a fibre nib; this type will produced a good clean trace. However, the fibre-nib pen is not recommended for a hygrograph as its weight introduces errors in the record which are difficult to assess. Particular care should therefore be taken to obtain a good record with a metal pen. Ensure that the pen rests as lightly as possible against the chart as is consistent with a good record.

8.6.3.1. *Thermograph.* Clean the instrument regularly, handling it carefully to avoid damage. Wipe the case and the base plate with a clean soft rag. Treat ink-stains promptly before they dry and harden. Clean any dirt from the bearings and apply, very sparingly, a little clock oil. Remove any surplus oil with blotting paper.

8.6.3.2. *Hygrograph.* Prior to any instrument maintenance on the hygrograph, the wire cage over the hairs should be removed and the hairs unhooked with a camel-hair brush. Clean the instrument and remove any ink-stains with methylated spirit or other suitable fluid. Clean the quadrants by rubbing the rolling surfaces with blotting paper which has previously been rubbed with a soft lead pencil. As with the thermograph, the bearings should be cleaned and very lightly oiled, any excess oil being removed with blotting paper. Great care must be taken to ensure that no oil gets on to the hairs. When replacing the hairs on to the hook the camel-hair brush must be used. This maintenance should be carried out once a month. At more frequent intervals the hairs should be washed. This should be done at least once a fortnight in country air but more frequently where the hairs are exposed to sea spray or industrial fumes and dust. Apply purified water to the hairs with a soft camel-hair brush. The wire cage must, of course, be removed for this purpose and great care taken not to touch the hairs with the fingers. After the excess water has been removed with the brush, the weight of water remaining is still quite appreciable for a time and this is enough to cause sagging of the hairs to the extent that the chart record will indicate about 95 per cent and not 100 per cent. This fact can be used when resetting the instrument. If it needs adjustment, move the jaw holding one end of the hairs and use the fine adjustment screw for which a key is provided.

8.7. ASPIRATED PSYCHROMETER

The values of humidity deduced from dry-bulb and wet-bulb thermometers exposed in standard screens are subject to uncertainty on account of variations in the rate of movement of air past the wet bulb. This uncertainty is generally ignored for ordinary synoptic or climatological purposes. When more precise values of humidity are required, or if atmospheric temperature and humidity readings are a requirement where there is no screen available or the installation of one would not be feasible, these readings can be obtained by the use of an aspirated psychrometer (Plate XXVII), also known as an aspirated hygrometer. The method of aspiration is by a fan, operated either by an electric or clockwork motor, which produces a fast and regular flow of air over the bulbs of two thermometers mounted in a frame. Each bulb is protected from external radiation by highly polished double-walled radiation shields.

Normally observations should be made with the instrument suspended from a fixture so that the bulbs of the thermometers are 1·25 m above the ground. The rod issued with the instrument may be screwed into any convenient object, preferably a fairly thin post away from other objects. If the support is of appreciable size, the instrument should be mounted on the windward side. If no convenient supports are near, the instrument may be held to windward at arm's length and tilted slightly so that the air inlets face approximately into the wind but not directly towards the sun. If the wind speed is greater than 12 knots and the instrument is hanging vertically from the support rod, inaccuracies in the readings may occur as the wind blows horizontally past the bottom of the radiation shield and disturbs the airflow over the thermometers. In that case the instrument should be hand-held instead, as described above.

The most important single factor affecting the accuracy of the readings is the wick of the wet-bulb thermometer. It should be a close-fitting piece of tubular wick of correct length; a loose-fitting or too short a wick may give rise to serious errors. When a new wick is received it will almost invariably be ready for use. If not, it should be treated as described in 8.4.1.1(b), page 119. In order to fit the wick, one of the polished radiation shields must be removed by giving it a short turn to the left to release the bayonet fitting. The piece of wick must be of such a length that it extends at least 12 mm above the top shoulder of the thermometer bulb. The wick should be handled as little as possible, and then only with clean fingers. If an instrument is used at only irregular intervals, a new wick should be fitted each time. If in regular use, the wick should be changed at least once a week; at sea or near large towns or in polluted areas it should be changed two or three times a week.

The following procedure should be followed when making an observation:

(a) The wick should be moistened by means of the injector provided. The injector is filled with purified water and the bulb pressed until the water rises to the top of the tube. The tube is then pushed over the wick and left there until it is completely saturated. The wick should never be moistened by using the injector as a squirt. If after saturation there is a residual drop of water on the bottom of the wick, this should be removed by touching the drop with the end of the tube. Care must be taken not to push the wick above the thermometer bulb when using the injector. If the wick becomes frayed at the bottom it should be changed.

The temperature of the purified water should be approximately the temperature of the wet bulb, otherwise a period of 1½ to 2 minutes may be required for the wet bulb to stabilize after each moistening. A wetting may only last about 5 minutes in conditions of low humidity. When the wet-bulb temperature is below freezing, ice-cold water should be used and, if necessary, the water should be induced to freeze on the wick by touching it with ice or hoar frost (as recommended in 8.4.1.1(e) for ice-bulb temperatures in screens.).

(b) Start the fan. One full winding of the clockwork motor should run the fan at the correct speed for at least 7 minutes. As soon as a slackening of the speed of the fan becomes noticeable through a lowering of the pitch of the note produced, the motor should be rewound. Readings taken at less than the full fan speed are liable to be in error.

(c) Two minutes after commencing aspiration, and again after a further half-minute, the thermometers should be read, wet bulb first and then the dry bulb. If these consecutive readings agree within ± 0.1 °C, note the actual readings of the wet- and dry-bulb thermometers. Sometimes consecutive readings may differ by as much as 0.5 °C and in these circumstances the mean of readings at half-minute intervals from minute 2 to minute 5 should be taken.

As stated in 8.2.1.1, thermometers used in psychrometers with forced ventilation are issued with certificates giving the corrections at different points on the scale; these must be applied before using the readings. Special scales for computing humidities from aspirated psychrometer readings are provided on the Meteorological Office humidity slide-rules, Mk 6 and Mk 6A. Tables for use with aspirated psychrometers are given in *Hygrometric tables*, Part III. Dew-points in degrees Celsius and tenths can also be obtained from special tables used by the Observational Practices Branch of the Meteorological Office.

8.7.1. Routine maintenance. The chromium-plated parts of the instrument should be rubbed from time to time with a soft cloth which is clean and dry; no metal polish should be applied. Care must be taken not to scratch the plating.

The bearings of the clockwork motor should be oiled occasionally with a little clock oil. To gain access to the motor, the winding key must first be removed by screwing counter-clockwise; if the six screws in the casing are then taken out, the casing will slide off, exposing the motor.

The thermometers should be cleaned periodically, especially when an instrument is used in a salt-laden atmosphere. Access can be gained by loosening the two locking nuts under the fan housing and detaching the frame from the fan. The thermometers can be released by turning the locking ring so that the recesses are over the thermometer apertures. The rubber spacing rings should be replaced as soon as they show sign of wear.

8.7.2. Safety precautions.   Particular attention must be paid to safety precautions when using a psychrometer with an electrically driven fan.

CHAPTER 9

PRECIPITATION

9.1. GENERAL REQUIREMENTS

The total amount of precipitation which reaches the ground in a stated period at any place is expressed as the depth to which it would cover a horizontal surface at that place if there were no loss by evaporation, percolation or run-off. The precipitation may be liquid (rain or drizzle) or frozen (snow, snow pellets, snow grains, hail, small hail, ice pellets, diamond dust) or a mixture (rain and snow, drizzle and snow, rain and melting snow). Precipitation is described as freezing rain or freezing drizzle when the drops of rain or drizzle have temperatures below 0 °C and freeze on impact with the ground or with objects on the earth's surface. Surface condensation phenomena such as dew, wet fog, hoar frost and rime may contribute to the catch of a rain-gauge but are not classed as precipitation under the heading of 'present weather'.

The total precipitation recorded and reported is the sum of the amount of liquid precipitation and the liquid equivalent of any solid precipitation (that is if precipitation falling as snow or ice were melted). Somewhat different methods of measurement are required according to the nature of the precipitation. The standard unit of measurement nowadays is millimetres and tenths, with depths of snow measured in centimetres.

9.2. BASIC METHODS OF MEASUREMENT

Bearing in mind that falls of precipitation are rarely either uniform in intensity or duration, the chief aim of any method of measurement of precipitation should be to obtain a sample that is truly representative of the fall over the area to which the measurement refers. Therefore choice of site, the design and exposure of the rain-gauge, the prevention of loss by evaporation and the effects of wind and splashing are important points to be observed.

The method of obtaining a sample of precipitation is for it to be collected in a rain-gauge which is sited in a representative spot. In its simplest form a rain-gauge consists of a circular collector, of known area, which must be horizontal and with a sharp upper rim. The precipitation falling within this collector is channelled through a funnel, either to a receiver for measurement later or to automatic measuring equipment. The rim of the collector can either be exposed at a standard height above ground or set in the centre of a specially constructed, large-diameter pit with its rim flush with the surrounding ground. Both methods of exposure have their disadvantages. The amount of rainfall collected by the conventional types of rain-gauge protruding above otherwise open terrain is generally less than the actual rainfall at the site of the gauge; this effect is mainly due to the eddies set up by the gauge itself when the wind is strong. A gauge standing in a pit with the rim of the gauge flush with the surrounding ground and with the pit covered by a gridded (or

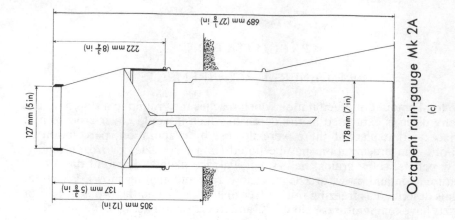

Octapent rain-gauge Mk 2A

(c)

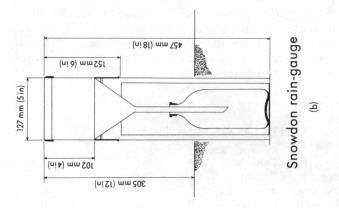

Snowdon rain-gauge

(b)

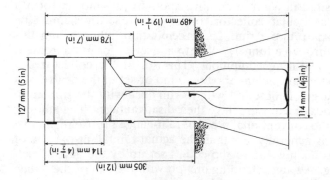

Meteorological Office
rain-gauge Mk 2

(a)

Figure 12.   Approved types of rain-gauges

other suitable artificial) surface to minimize the effect of in-splash into the gauge itself may also be affected by wind eddies and can suffer drainage problems.

Comparability of readings within a rain-gauge network can best be achieved if the measurements are all made with an approved type of instrument in a standard exposure. The Meteorological Office has a strong preference for the gauges described in 9.3, that is for those in which the rim is 5 inches (127 mm) in diameter and sited at 30 centimetres (12 inches) above the ground (rather than flush with it). The exposure must also be standard, and determined by the rules given in I.9 of Appendix I (page 188).

The collection of rain and drizzle by a rain-gauge, and its subsequent measurement, is simple. Hail or sleet may be similarly sampled and collected within the rain-gauge and, after thawing by natural or artificial means, the liquid equivalent may be measured.

Measurement of the liquid equivalent of snow presents the difficult problem of securing a representative sample, and no really satisfactory method is yet available. In light winds, falls of snow collected in the rain-gauge and not exceeding its retaining capacity should be thawed and then measured as liquid water. For heavier falls, and for falls in stronger winds, the method that offers the best guidance is the collection of samples of fresh snow of known horizontal cross-section that has accumulated upon the ground since the previous specified time of measurement. These samples are subsequently measured by weight or, when completely melted, by volume.

9.3. MEASURING PRECIPITATION BY COLLECTION

Many different types of rain-gauge have been designed and used. All consist of a circular collector, delineating the area of the sample, and a funnel leading into a reservoir. The precipitation collected is then measured directly in millimetres and tenths using a rain measure graduated for use with the particular type of gauge. The shape and size of gauges vary according to the storage capacity required in the reservoir, this being dependent on whether the precipitation is measured daily, weekly or monthly.

9.3.1. The Mk 2 rain-gauge. The standard rain-gauge used in the Meteorological Office for climatological observations is the Mk 2 (Figure 12(a)). This is a copper gauge with a 5-inch diameter (127 mm) collector which has a sharp brass rim. It is sited so that the rim is horizontal and 30 cm (12 inches) above the ground. The rainfall that is collected is led into a narrow-necked bottle placed in a removable copper can. The bottle is easy to handle when pouring the sample into the rain measure, and reduces loss by evaporation. The inner can ensures the retention, for measurement, of exceptionally heavy rainfall which may overflow from the bottle into the can. Both bottle and inner can are housed within the splayed base of the gauge which is sunk firmly into the ground. An additional 5-inch gauge may be provided at those stations which report rainfall for synoptic purposes.

9.3.2. The Mk 3 rain-gauge. This rain-gauge is of glass-fibre construction. It was originally designed as a 'system' which could act either as a recording rain-gauge (see 9.5.2) or as a collector. In its latter capacity it can be fitted with a circular collecting funnel of 127 mm diameter and, when exposed, the flared support tube must be buried so that the rim of the collector is 30 cm above ground level and horizontal. This gauge is not recommended for use as a manually read instrument because it is brittle and the collector is liable to chip and crack.

9.3.3. The Snowdon, Octapent and Bradford rain-gauges.
Another pattern of rain-gauge which is used when daily readings are required is the Snowdon gauge, shown in Figure 12(b). However, this particular gauge does not have a splayed base and it is not so firm in the ground as the Mk 2 gauge.

The Octapent rain-gauge (Figure 12(c)) provides sufficient capacity for periods up to a month or more between measurements. A smaller capacity Octapent (680 mm) is used in the drier parts of the country and a larger capacity gauge (1270 mm) in areas of high rainfall. A special design feature of the Octapent rain-gauge is the frost protector. This is a piece of reinforced rubber tubing, closed and weighted at the lower end, which is inserted vertically in a hole set to one side of the top of the inner container. The protector operates by collapsing under the pressure due to expansion when the contents of the inner container freeze; when a frost protector is not used, the vacant hole should be sealed to reduce evaporation losses.

The Bradford rain-gauge (not illustrated) is an adaption of the Snowdon gauge with a deeper storage can to hold the rainfall. This gauge is capable of being read monthly in fairly dry localities but is more often used for weekly readings.

The measurement of the rainfall in the Octapent and Bradford gauges is done in two stages. As a crude check on the later accurate measurement, the rain is first roughly measured by inserting a graduated dip-rod which is lowered vertically until its metal foot touches the bottom. The rod, when withdrawn, is wet up to the mark indicating the amount of rain. This amount is noted and the rain is then measured accurately by means of a glass measure in the usual way. If fitted, the frost protector in the Octapent gauge should be removed before a measurement is made with the dip-rod.

9.3.4. Rain measures.
The amount of precipitation collected by a gauge may be measured with the aid of glass vessels known as rain measures. Each measure is graduated to indicate, in millimetres, the amount of precipitation over the area of the collecting funnel of the rain-gauge; it is therefore essential that the rain measure should be one appropriate to, and calibrated for, the type of rain-gauge in use.

Meteorological Office rain measures (see Figure 13) for use with daily gauges are tapered at the base to facilitate accurate measurements of small amounts. There are two standard rain measures in use, the older one having a total capacity of 10·0 mm to the highest graduation and the newer having a graduated capacity of 10·5 mm. Amounts greater than the graduated capacity must therefore be measured in two or more stages. It is not necessary to fill the measure exactly to the highest graduation. For example, a rainfall of 30·8 mm might be summed in four stages:

$$9·6 + 9·4 + 9·5 + 2·3 = 30·8 \text{ mm}$$

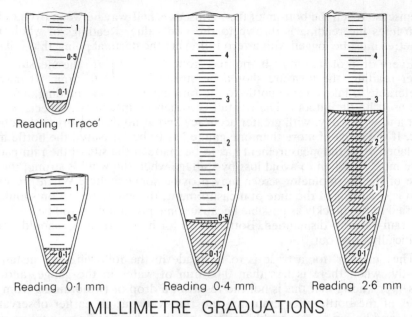

MILLIMETRE GRADUATIONS

Figure 13.   Reading the rain measure

The reading 'trace' is also applicable when rain is known to have fallen but the gauge is dry
(see page 140)

9.3.4.1. *Reading the measure.*   Owing to the surface tension, the surface of
the water in a rain measure is never quite flat but is slightly curved, the water
being drawn up at the sides as illustrated in Figure 13. The rule is always to
read the lowest point, which is the centre of the meniscus, because measures
are graduated on this basis. To avoid errors of parallax the observer must
hold the measure vertical with the meniscus exactly at eye-level when making
a reading. The tapered measure is provided with duplicate engravings of the
main graduation marks at the back of the glass to facilitate accurate reading.

The larger labelled divisions on the glass represent whole millimetres and
the intermediate divisions represent tenths. The lowest division, the marking
of which is carried completely round the glass, is figured 05 on some measures
and represents half a tenth of a millimetre (0·05 mm). It is there for a special
purpose, namely to distinguish between a reading to be logged as 0·1 mm and
a reading to be logged as a 'trace'. If the bottom of the meniscus is below this
special graduation the entry is trace (abbreviated in the Register to 'tr'). If the
bottom of the meniscus is exactly on the special graduation, or between it and
the next division, the reading is logged as 0·1 mm.

Higher readings are expressed to the nearest tenth of a millimetre, thus 0·7,
1·0, 2·3, etc. The figure '0' must be entered before the decimal point when the
amount is below one whole millimetre. In other instances the whole millimetres are given by the number against the figured main division next below
the meniscus, and the tenths are obtained by counting upwards from this
division to the intermediate division nearest in line with the bottom of the

meniscus. When the bottom of the meniscus is half-way between two graduation lines the reading is thrown to the odd value. Readings of whole millimetres must be logged with a zero following the decimal point, thus 2·0 etc.

Every drop of rain must be poured from the bottle into the measure and,
after reading, the measure should be stored inverted to drain. It may be
preferable to take a clean bottle to the rain-gauge at every reading, and take
the used bottle indoors. The rainfall can then be measured, especially on a
wet and windy day, with greater accuracy and at no discomfort to the observer. If the catch of more than one gauge has to be measured the bottle must
be labelled. If the measurement has to be made at the site of the rain-gauge,
care must be taken to avoid loss by spillage when the wind is strong; the lee
side of the thermometer screen may provide some shelter. When precipitation is occurring at the time of measurement, the whole operation should be
carried out as quickly as possible because some precipitation will be lost while
the rain-gauge is dismantled. Bottles used for hourly readings should not be
artificially dried out.

The entry 'tr' for a trace is to be made in the following two instances.
Firstly, when there is less than 0·05 mm of water in the gauge, and the
observer knows that this is not the result of a drop or two draining from the
sides of the bottle after emptying the rain-water at an earlier observation
time, and he can be reasonably certain that there has been precipitation since
the preceding measurement. If the observer knows that the deposit in the
gauge results from dew, wet fog, hoar frost or rime, an appropriate note
should be made in the remarks column of the register (for example, tr(w),
tr(fe), tr(x)). (The entry tr(w) is not made just because the observer sees dew
on the grass.) Such deposits are sometimes more than 0·05 mm, in which case
the measured amount is recorded. Secondly, the entry is to be made when the
observer knows definitely from his own observation that precipitation (other
than dew, wet fog, hoar frost or rime) has fallen since the preceding observation and yet finds no water in the gauge. This happens sometimes, especially
in warm dry weather; the gauge may not even be damp, the small amount of
rain having evaporated before it could run into the bottle. In such circumstances, particularly when no scheduled observation is made at a time which
would include note of the precipitation, the observer should enter an
appropriate note such as 'slight shower of rain' (or the appropriate Beaufort
letters) and the time, if known, in the remarks column of the Register.

When the amount of rainfall is nil, the entry in the Register should be '—'
not '0·0', but for certain data sheets different instructions may be given.

9.3.5. Care of the rain-gauge and rain measure. The gauge should be inspected from time to time to ensure that it is in sound condition and free from
leaks. Such inspections are specially important for copper gauges which may
become distorted or develop leaks. If a Mk 3 gauge is being used it should be
noted that this type, made of glass fibre, is liable to chipping of the rim and
cracking of the collector.

The best method of testing the funnel is to close the outlet tube with the
thumb and plunge the funnel, rim downwards, into a bucket of water; any
leak will immediately be revealed by air bubbles. Water should never be
found in the outer case unless the bottle and inner can of the copper gauge

have overflowed after quite exceptional rainfall. If water is found in the outer case under any other circumstances the case must be removed from the ground, tested for leaks, and any leaks repaired before it is set properly back into the ground.

The height and level of the rim of the gauge should be checked from time to time and corrected if necessary.

The gauge must be kept clear of fallen leaves or other debris which might block the opening and thus prevent the collected water from flowing into the receiver.

Bottles used with gauges should be kept clean; cracked bottles should be replaced at once. Beside being narrow-necked, the bottle should be sufficient-ly tall to permit the delivery tube to enter it, yet not too tall to foul the underside of the funnel. The delivery tube must be guided gently into the neck of the bottle. If the tube is banged hard against the bottle it will cause ridging and denting at the base of the funnel. The measure should also be kept clean and should be stored, preferably inverted, in a safe place when not in use.

Care must be taken to avoid damage to the rain-gauge when the grass is being cut around it. To reduce the risk of the gauge being damaged by grass-cutting equipment the turf should be removed for a few centimetres around the gauge and replaced by small stones or granite chippings; never surround the gauge with concrete, tarmac or other solid surface as this may cause insplashing.

9.3.6. Hours of reading. All types of station equipped with rain-gauges should take readings at 0900 GMT. Additional readings are taken at certain types of station in the United Kingdom as follows:

(a) Meteorological Office and auxiliary stations which make observations at 2100 GMT should take additional measurements at that time.

(b) Stations which contribute observations to the Health Resort Bulletin should take additional measurements at 6 p.m. clock time.

(c) All Meteorological Office stations, and those auxiliary stations making full synoptic observations and are able to, should take additional measurements at 0000, 0600, 1200 and 1800 GMT for synoptic pur-poses.

To ensure as far as possible the accuracy of rainfall measurements for climatological purposes, the standard rain-gauge should be used only for measurements at 0900 and 2100 GMT; stations as in (b) above may also use the gauge for their special readings at 6 p.m. clock time, pouring the water back into the collector when the measurement has been made. Rainfall totals should be carefully checked to ensure that they refer to the correct overall period when intermediate measurements are taken. Meteorological Office and auxiliary stations which make additional readings at other times should use a second rain-gauge for this purpose; this second gauge will be either a tipping-bucket or a 5-inch gauge or both (the latter acting as a back-up to the former where both are provided). This second gauge (or gauges) should be sited in the most convenient position which satisfies the rules for rain-gauge exposure (see I.9 of Appendix I, page 188).

Special instructions are issued to stations abroad.

9.4. MEASUREMENT OF SOLID PRECIPITATION

Special methods of measurement are required when the rain-gauge contains solid precipitation, whether originating from snow, snow pellets, snow grains, hail, small hail, ice pellets, diamond dust, or a mixture of rain and snow (sleet), or from rain-water which has frozen after falling.

The World Meteorological Organization has laid down that 'the amount of precipitation shall be the sum of the amounts of liquid precipitation and the liquid equivalent of solid precipitation'. With slight snowfalls, in little or no wind, no especial difficulty is encountered and the procedure in 9.4.1 should be followed. Reliable measurements of snowfall in stronger winds are very difficult and depend much on the zeal and skill of the observer in following the guidance given in 9.4.2.

The rainfall equivalent of any snowfall is determined in two ways, depending on the circumstances at the time of the snowfall. These are either the measurement of the liquid equivalent of any snow which has accumulated in the gauge funnel and receiver since the previous specified time of measurement or by obtaining an estimate of the equivalent rainfall by sampling level undrifted fresh snow. A quite separate climatological observation of the total depth of snow lying is made at 0900 GMT (see 9.7). There are a number of important reasons for requiring these measurements; not least is their use in the assessment of the amount of water which would be released in a sudden thaw, and the possibilities of river flooding.

In discussing the two measurements required when snow has fallen, namely the liquid equivalent and total depth of snow lying, reference is made to level undrifted snow. It is important, particularly during any prolonged cold spell when there may be further falls of fresh snow, that the area where measurements are made is as free from disturbance as is practicable. Random trampling of the snow surface may make it difficult to obtain representative samples. The path to and from the area should be chosen with care and maintained on subsequent visits, extending for further sampling only as necessity dictates.

9.4.1. Measurement of solid precipitation using the rain-gauge: liquid equivalent. When there have been no more than light winds during the period since the previous scheduled observation and there has been a comparatively slight fall of solid precipitation, the amount (liquid equivalent) may be measured by one of the following three methods; the first method should be used only when precipitation is not occurring at the time of observation.

(*a*) The rain-gauge funnel and receiver are brought indoors, the contents melted and the water measured in the normal way. The top of the funnel should be covered by a flat plate to minimize the loss by evaporation. Excessive heat should not be applied because, in addition to the possible loss of precipitation by evaporation even if the funnel is covered, there is a risk of melting the solder on a copper funnel.

(*b*) A cloth dipped in hot water is applied to the outside of the funnel and receiver to melt the snow or ice. Ensure that no water from the cloth enters the gauge.

(*c*) Some warm water is accurately measured in the appropriate rain measure and then poured into the gauge. Only sufficient warm water to melt the snow or ice should be used. The amount of water added is subtracted from the total measured. About two rain measures full of water at about 40 °C will be required if the funnel is full of snow. This method is to be preferred for glass-fibre gauges.

When rain collected in the bottle turns to ice, the bottle must be warmed gently to thaw the ice without cracking the bottle. If the bottle is found to be cracked the ice must be thawed in such a way that no part of the catch is lost. The observer must take care to avoid being cut by broken glass.

9.4.2. Measurement of solid precipitation: other methods. Moderate or heavy falls of snow, particularly if accompanied by other than light winds, will probably make the methods described in 9.4.1 inappropriate. Wind eddies may carry snow clear of the gauge or even out of the collector. If the wind is strong, lying snow may be raised into the air and be blown into the gauge. In extreme cases the entire gauge may be buried in snow. In those circumstances the rain-gauge is unusable and special methods of measurement are required. The procedures which should be adopted are detailed below.

9.4.2.1. *When all the precipitation has occurred as snow and no snow was on the ground prior to the period of snowfall to be measured.* Obtain a sample by pressing the inverted funnel of the rain-gauge (not a glass-fibre gauge) vertically downwards through level undrifted snow until it reaches the ground. Melt the sample indoors and measure the liquid in the rain measure appropriate to the rain-gauge (to the nearest 0·1 mm).

It can be extremely difficult to judge from an extensive stretch of apparently level and uniform snow where a representative sample should be taken. For this reason it is advisable to take three samples several metres apart. Each of the three samples should be melted separately and the average reported. The maximum and minimum values of the liquid equivalents should be noted so that, in the event of the rainfall equivalent being queried, some indication can be given of the possible spread of the readings upon which the final reported figure will have been based.

When depths greater than about 15 cm occur, each sample should be taken by dividing the total depth into two or more layers by inserting a piece of sheet metal or hardboard into the snow at the appropriate depth, after clearing some snow adjacent to one side of the sampling area.

An alternative method is available for those observers who have accurate weighing facilities. A tube of about 5 to 8 cm diameter and with a sharp edge is used to obtain a core of snow by pressing the tube vertically downwards through level undrifted snow until it touches the ground. Weigh the tube and the sample and, after deducting the weight of the tube, convert the weight of the snow to an equivalent depth of rainfall, as in the following example.

$$\text{Weight of tube + snow} = 55\cdot1 \text{ grams}$$
$$\text{Weight of tube} = 35\cdot0 \text{ grams}$$
Thus weight of sample collected $= 20\cdot1$ grams.

The 20·1 grams of snow equates to 20·1 grams of water which has a volume of 20·1 cm³. The rainfall equivalent of the snowfall collected is given by

$$\frac{20\cdot1}{\text{Cross-sectional area of the tube}} \text{ cm.}$$

If the internal diameter of the tube is 5·0 cm, its cross-section area is $\frac{\pi \times 5 \times 5}{4}$ cm², i.e. 19·6 cm².

Thus the rainfall equivalent becomes $\frac{20·1}{19·6}$ cm = 1·03 cm = 10·3 mm.

The tube method does not work with wet snow.

9.4.2.2. *When snow has fallen or been blown on to the ground prior to the period of snowfall to be measured*, a clean surface must first be prepared to receive any further snowfalls. This can be provided by a wooden board (with a slightly rough surface, painted white) placed level on the top of the snow with its upper surface flush with the snow surface. The board should be swept clean of snow immediately after each snowfall measurement and replaced on the level snow as before.

The selected site should avoid places where drifting is likely or over-exposed places where snow is unlikely to accumulate in strong winds. The position of the board should be indicated by a thin cane so that it can be located under the snow.

When such a board is used, samples of snow taken from it can be treated by either of the methods in 9.4.2.1 to obtain the liquid equivalent.

9.4.3. When precipitation changes from snow to rain or drizzle between scheduled hours of observation. No special action is required when conditions described in 9.4.1 hold. In other conditions the observer should, as soon as possible, carry out the procedure detailed in 9.4.2.1 or 9.4.2.2, as appropriate, and set the rain-gauge ready to receive the liquid precipitation then falling. A note should be made in the Register of the time and the liquid equivalent of the sample taken. This liquid equivalent should be added to the amount of precipitation subsequently caught in the normal way by the rain-gauge to give the total amount of precipitation at the next scheduled hour of observation.

9.5. RECORDING RAIN-GAUGES

Recording rain-gauges are used to keep a continuous record of rainfall amount against time. The type principally used by the Meteorological Office is the tilting-siphon rain recorder (see 9.5.1). A tipping-bucket rain-gauge (see 9.5.2) can be converted to a rain recorder by the addition of a recording unit (see 9.5.3.).

Rain recorders are usually installed in conjunction with a standard collecting rain-gauge. It is the measurement of the rain in the standard rain-gauge which is reported for climatological purposes. At those selected stations which are issued with a tilting-siphon rain recorder the total rainfall from the standard gauge is supplemented by the analysis of the permanent record chart to provide details of times of onset and cessation of rainfall and any variations of intensity. A tipping-bucket gauge is usually used when rainfall is reported for synoptic purposes (although manually read gauges are also used), and the rainfall derived from the number of tips over a period (hourly or six-hourly) is the amount reported in this particular case. With the introduction of the magnetic-tape event recorder (see 9.5.3), subsequent interpretation of the magnetic tape can provide similar information to that from the tilting-siphon rain recorder.

Details of several types of recording rain-gauges are given in the *Handbook of meteorological instruments*.

9.5.1. Tilting-siphon rain recorder. The mechanism of the Meteorological Office tilting-siphon rain recorder, Mk 2, is illustrated diagrammatically in Figure 14. The water collected by a large funnel 287 mm (11·30 inches) in diameter falls into a chamber A and raises a plastic float B which is provided with a vertical rod to which the recording pen is attached. The chamber A is mounted on knife-edges C and is so arranged that it overbalances when full of water, sending a surge of water through the double siphon tubes D which empty the chamber and allow the pen to return to the zero of the chart. The overbalancing is controlled by a trigger E which is released by the rising float at a predetermined point. As the float chamber empties, the centre of gravity shifts to the left and under the action of the counterweight F the system resets itself to the normal working position. A rain trap G reduces the loss of precipitation during the siphoning period which should be completed within 8 seconds. By means of an upright rod H, the pen is automatically lifted clear of the chart as soon as the chamber starts to overbalance.

The earlier version of this recorder, the Mk 1 (Plate XXVIII), has only one siphoning tube and in consequence the siphoning takes a minimum of 15 seconds. The float, being of metal, is liable to sustain damage if the water in the chamber becomes frozen. There is no rain trap to reduce the loss of precipitation during the siphoning period.

The metal float cannot be interchanged with the plastic float but the whole of the float-chamber assembly can be replaced by the Mk 2 float-chamber assembly. This assembly not only includes the plastic float but also the two siphoning tubes and the rain trap. Full details and instructions are given in the instructions which are available from the Operational Instrumentation Branch of the Meteorological Office.

9.5.1.1. *Method of use.* The chart should be changed daily at about 0900 GMT and as nearly as possible at the time of reading the standard rain-gauge; at the same time the clock should be wound. The pen reservoir should be topped up if a fibre-tipped pen is not in use. When fitting the new chart on the drum, corresponding graduations on the overlapping portions should be coincident and the chart should be as close as possible to the flange on the bottom of the drum. The end of the chart should overlap the beginning. When adjusting the drum so that the pen is at the correct time on the chart, the drum should be turned counter-clockwise (when viewed from above) to avoid backlash, but every effort should be made to set the time correctly at the first attempt. Only daily clocks may be used.

A time mark should be made on the chart within two hours of the commencement of the record (but not within the first half hour) by gently depressing the float-chamber centre spindle enough to make a small mark. If possible another time mark should be made when the record is about half completed. Care should be taken that the pen does not 'bounce' above the rainfall trace. No time mark should be made when precipitation is occurring.

Full station details, the times of the time marks, the time the record commenced and ended, should be entered on the old chart. The times should be noted to the nearest minute. The serial number should be taken from Met. O. Leaflet No. 11, The Meteorological Office Calendar.

The readings of the total rainfall each day should always be compared with the readings of a standard rain-gauge exposed nearby. Any errors in the readings of the recording instrument, apart from small discrepancies caused by slight under-recording of the value of total precipitation, should be investigated immediately.

9.5.1.2. *Testing the tilting-siphon rain recorder.* The instrument should be tested regularly, especially when no precipitation has occurred for some time, by pouring successively into the funnel equal quantities of water corresponding to some specific interval on the chart. This quantity is best measured out in a rain measure, but the equivalent rainfall has to be calculated by taking into account the relative diameters of the apertures of the tilting-siphon gauge and the gauge for which the measure was graduated. For example, if a rain measure graduated for a 5-inch rain-gauge is used, the amount in the measure needed to obtain a reading of one millimetre on the tilting-siphon would be $(11 \cdot 31/5)^2 = 5 \cdot 1$ mm.

9.5.1.3. *Care of the tilting-siphon rain recorder.* The inside of the funnel should be kept clean by being rubbed with a damp rag; polish must not be used. The hinge should be oiled occasionally.

The time taken to siphon should be checked occasionally to establish that the outlet tubes are not partially blocked; the correct siphoning period should be about 8 seconds for a Mk 2 recorder and 15 seconds for the Mk 1. The drain-filter through which water from the siphon tube passes to the base of the instrument should occasionally be inspected and cleaned if possible, although access to this part is very difficult. Methylated spirit may be used if necessary.

The rain trap fitted above the float chamber should periodically be taken off and inspected and cleaned. The filter should be cleaned with methylated spirit. If the small hole in the rain trap leading to the drain becomes blocked, the small rubber cap on the end of the tube should be removed and the hole cleaned out with a piece of wire. A light smear of silicone grease applied very occasionally to the inside of the rain trap will improve the run-off, but care should be taken that the grease does not block the hole.

9.5.1.4. *Frost protection.* In temperatures below freezing, if no form of heating has been installed, the water in the float chamber will freeze, thus causing the recorder to become inoperative. The plastic float in the Mk 2 recorder will not be affected by the ice in the float chamber but the metal float in the Mk 1 will be liable to damage. To ensure continuous efficient operation of the recorder in frosty weather some form of heating is required. This heat is not intended to melt frozen precipitation but purely to prevent freezing of the rain-water already inside the chamber of the gauge.

The Meteorological Office uses a standard rain-gauge heater in which a low-voltage, low-wattage, flexible heating element is used in conjunction with a thermostat. The heating element operates from a 24-volt d.c. or a.c. supply, and either heavy-duty accumulators, or a mains supply through a suitable

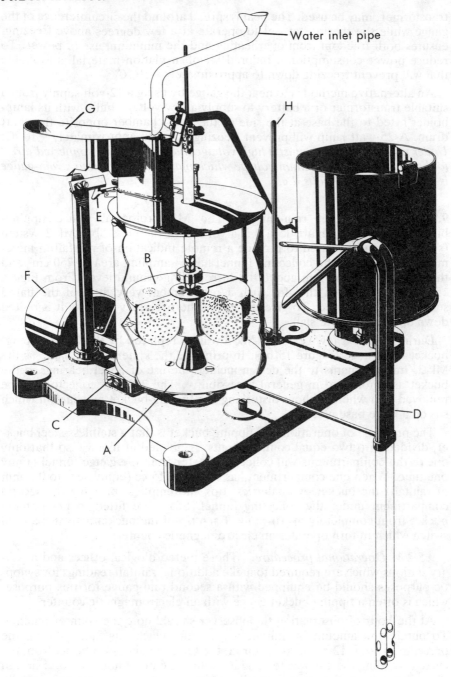

Figure 14. Meteorological Office tilting-siphon
rain recorder

A	Collecting chamber	E	Trigger
B	Plastic float	F	Counterweight
C	Knife-edges	G	Rain trap
D	Double siphon tubes	H	Pen-lifting rod

transformer, may be used. The heat is spread around the circumference of the gauge while the thermostat, set to operate at a few degrees above freezing, ensures both freedom from overheating and the minimum use of power. To reduce power consumption a tailored kit of insulation material is available that will prevent freezing down to approximately −10 °C.

An alternative method is to heat the gauge by using a 12-volt supply from a suitable transformer or a battery to supply a low-voltage bulb, with its lamp-holder fitted to the base at the side of the float chamber opposite the chart drum. A 25-watt lamp will prevent freezing down to approximately −10 °C. *Under no circumstances must a high-voltage mains supply be connected direct to the recorder; the maximum voltage which may be connected to a rain-gauge is 30 volts a.c. or 50 volts d.c.*

9.5.2. Tipping-bucket rain-gauges. The Meteorological Office tipping-bucket rain-gauge (Figure 15(a)) is a version of the glass-fibre Mk 3 system (see 9.3.2) which is used to provide a remote indication of rainfall in incre-ments of 0·2 mm. The collecting funnel has a sampling area of 750 cm² and the rim must be 450 mm above the surrounding ground level. From Figure 15(a) it will be seen that for such a requirement very little of the flared support tube is buried. To give the gauge the necessary stability it is bolted down on to a concrete slab.

During the next few years the Mk 3 will be replaced by the Mk 5 tipping-bucket rain-gauge (Figure 15(b)). In principle the system is the same as the Mk 3. Improvements to the design include the use of jewel bearings in the bucket assembly and in general accessibility. The 750 cm² collector can be removed as a whole without disturbing the tipping-bucket mechanism which sits on a large base plate.

The principle of operation of a tipping bucket is that a stainless-steel buck-et, divided into two equal compartments, is pivoted at its base so that only one of the compartments will collect the water from the gauge funnel at any one time. When one compartment has collected 15 cc (equivalent to 0·2 mm of rainfall) the bucket overbalances, tips and empties, bringing the second compartment under the collecting funnel. Stops are fitted to prevent the bucket from completely overturning. Each tip of the bucket activates a reed switch which in turn operates an electromagnetic counter.

9.5.2.1. *Operational procedure.* Those meteorological offices and auxili-ary stations which are required to make additional rainfall readings for synop-tic purposes should be equipped with a second rain-gauge for this purpose, which is often a tipping-bucket gauge with an electromagnetic counter.

At the hour of observation the observer should note the counter reading. To obtain the amount in millimetres of rain which has fallen during the preceding 1, 6 or 12 hours as required, the observer should subtract from the current reading the counter reading he noted 1, 6 or 12 hours before; he can then either multiply the result (the difference in the number of tips) by 0·2 mm or divide the difference by five to obtain the rainfall in millimetres; for example, 21 tips = 21 × 0·2 mm = 4·2 mm, or 21 ÷ 5 = 4·2 mm. If a small amount of precipitation is known to have occurred but there is no change shown on the counter, the observer is to record 'tr'; on certain data sheets an encoded entry representing 'trace' is made.

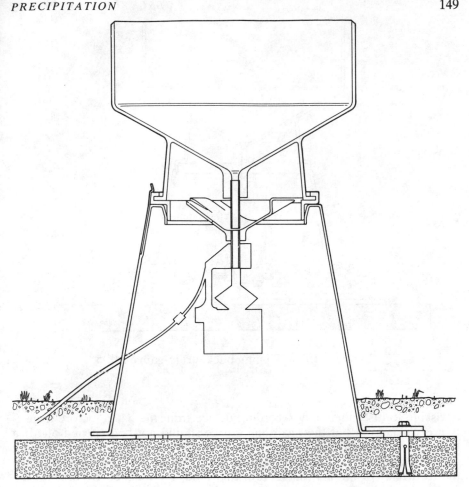

Figure 15(a).   Tipping bucket rain-gauge Mk 3

The procedure for dealing with solid precipitation in a tipping-bucket gauge is detailed in 9.5.2.5.

9.5.2.2. *Care of the tipping-bucket gauge.*   Once the rain-gauge is installed and calibrated, care should be taken not to disturb the system; for instance, a hard knock may alter the calibration or cause a nearly full bucket to tip.

The collector funnel must be kept free of obstructions such as fallen leaves, grass cuttings, etc., but any cleaning should be done gently without disturbing the tipping-bucket switch. Should it be necessary to detach the collector funnel of a Mk 3 gauge, it should be eased free of the retaining lugs without shaking the support tube and laid gently on its side to avoid chipping the rim. The funnel should be refitted to the support tube with equal care.

At six-monthly intervals the tipping-bucket switch should be carefully lifted just free of the gauge and the buckets examined to see if they contain silt. It is

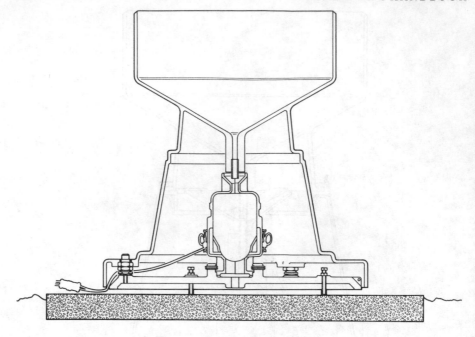

Figure 15(b).   Tipping bucket rain-gauge Mk 5

advisable to disconnect any recording device from the gauge before inspection is carried out but the recording device should not be switched off. If a bucket contains more than a light deposit of silt it should be cleaned by using a small paintbrush. Care must be taken during this operation not to damage the reed relay or the light metal arm that holds the magnet. If possible, the funnel should be replaced in the same position that it occupied before cleaning. Meteorological Office recorders will be serviced by the Maintenance Organization.

9.5.2.3. *Routine performance checks*.   At about monthly intervals, or when a cumulative rainfall total exceeding 10 mm has been reached, the total obtained from the standard check-gauge should be compared with the product of the calibration value and the number of tips obtained from the counter in the same period. Differences should not normally exceed 5 per cent. A discrepancy greater than 10 per cent indicates that the calibration has probably changed.

A simple test may be performed by slowly pouring water into the gauge and listening for the noise made when the bucket tips; the amount of water used must be noted and an appropriate correction made to the amount next reported. If no tips are heard, there may be excessive bearing friction or a leak between the buckets may allow the water to escape. If tips are heard but nothing is recorded a technician's services will be required.

9.5.2.4. *Calibration*.   The tipping-bucket rain-gauge should be calibrated at six-monthly intervals and at Meteorological Office stations this is carried out by the Maintenance Organization. The Operational Instrumentation Branch may be able to supply instructions, on request, for privately owned gauges.

9.5.2.5. *Solid precipitation in a tipping-bucket rain-gauge.* If snow or hail has fallen and there is solid precipitation in the gauge, hourly measurements are to be discontinued. The hourly entries over this period will be 'snow (or hail) affecting rain-gauge'. At main synoptic hours the funnel should be cleared of snow or hail (measurement of the liquid equivalent being made). This liquid equivalent is poured carefully back into the tipping gauge so as to obtain a final reading on the counter; this reading should be entered in brackets in the Daily Register.

9.5.3. Magnetic-tape event recorder. The Meteorological Office magnetic-tape event recorder is designed for use in remote locations with the tipping-bucket rain-gauge to provide a record of rainfall over periods of up to three months.

The battery-powered recorder consists of an incremental tape deck which records on magnetic tape contained in a large cassette. The recorder places a magnetic pulse on two of the four tracks each time the bucket tips, and a similar pulse is written on the remaining tracks once per minute following a signal from the recorder's internal electronic clock. Connections to the recorder should only be by circuits which have been specifically approved.

The recorder is housed in a hermetically sealed glass-fibre box and may be installed indoors or outdoors. Installation, operation and maintenance instructions are given in the handbook issued with the equipment. At Meteorological Office stations installation and regular maintenance is carried out by staff of the Maintenance Organization.

Although designed to run for longer periods, the cassette is usually changed at monthly intervals. The procedure for removing the used cassette and replacing it with a new one are dealt with fully in the operating instructions. The used cassette is sent to Meteorological Office Headquarters at Bracknell where the tape is translated and the information processed by computer to give hourly, daily and monthly rainfall totals. If the recorder unit is located outside and is brought inside a building for the cassette to be changed, care should be taken to prevent condensation forming on the inside of the recorder unit when it is opened.

9.6. GRAVIMETRIC RAIN-GAUGE

It has long been appreciated that the amount of rainfall collected by a conventional type of rain-gauge is likely to be less than the actual rainfall at the site of the gauge. The defect can arise from various causes, but by far the most common is due to the effects of the turbulent air flow near the ground. Various methods have been employed to eliminate or reduce the effect of the wind on the catch of a gauge; one method is to mount the gauge with its rim level with the ground (flush gauge) and to surround it with an artificial surface to minimize the effect of in-splash. Gauges mounted like this, together with their surrounding surface, can however still cause wind eddies which affect the catch.

The Meteorological Office has developed a flush type of gauge called a gravimetric rain-gauge which collects the precipitation in a pan of diameter 1·21 m mounted on a weighing-machine placed in a concrete-lined pit in the ground. The rim of the pan is level with the surrounding ground. The whole pan is covered by a stainless-steel mesh; the mesh and a circular area extending 2 m from the rain-gauge pan are covered with granite chips to form a homogeneous surface (apart from the small gap between the pan and the pit lining). The water can flow readily through the layer of granite chips into the pan. This construction of the collecting area of the gauge, together with its surrounds, has been designed to affect the airflow over the instrument as little as possible, and also to minimize the net effect of splashing. The weight of the pan and its contents is converted to give a millivolt analogue output which is registered on a recorder. Experiments have shown that this gauge will measure precipitation more accurately than other gauges used by the Meteorological Office but it is still undergoing tests; it is not anticipated that their distribution will be widespread when eventually they are introduced operationally. They will act as reference gauges at a few selected sites in the British Isles.

9.7. MEASUREMENT OF SNOW DEPTH

At climatological stations the total depth of snow lying at 0900 GMT is to be noted in the remarks column of Metform 3100 (Pocket Register). The symbol ⯈*⯈ may be used as an abbreviated indicator with the total depth shown at the right-hand side. The depth should be measured in centimetres, using a ruler* held vertically when more than half the ground representative of the station is covered in level snow, in a space free from drifting and not scoured by wind. The spot so described should be chosen to be as near as possible to the rain-gauge. If practicable, the mean of three such measurements at different spots should be taken.

At full synoptic stations the total depth of snow is reported at 0600 GMT in the coded synoptic message. In addition, the total depth of snow is reported in a special group at 0600, 0900, 1800 and 2100 GMT whenever snow is lying to an undrifted depth of 0·5 cm or more. If as a result of fresh snowfall, or the thawing of existing snow cover, the depth has changed by 2 cm or more since the last report, the new depth is reported at any main or intermediate synoptic hour. Most synoptic stations also report the total depth of snow and the depth of fresh snow at 0900 GMT in a separate message for climatological purposes. Instructions for observations of cover, character and regularity of snow at synoptic stations are given in the *Handbook of weather messages*, Parts II and III, and specific instructions are issued to certain stations.

---

*The ruler should either be adapted to read zero at ground level or alternatively a note should be made of the length of the short gap between the end of the ruler and the zero mark. A metre rule is preferable.

CHAPTER 10

SUNSHINE

10.1 GENERAL

The routine measurement of the recorded duration of bright sunshine, either in hourly or daily totals, is required from some stations for climatological purposes. The type of sunshine recorder approved and used by the Meteorological Office utilizes the heat from the sun's rays, focused by a solid glass sphere to an intense spot, to char a trace on a specially manufactured card. This was originally known as the Campbell–Stokes recorder.

For accuracy in both recording and measuring sunshine duration from an exposed card it is necessary for sunshine recorder to be correctly installed and adjusted. Full details are given in I.10 of Appendix I (page 190). It is essential that the recorder should be mounted on a rigid support, such as a concrete or brick pillar, so that any necessary adjustments may be considered reasonably permanent.

The Meteorological Office sunshine recorder Mk 2 was designed for use in latitudes between 45° and 65°, north or south. It has gradually been superseded by the Mk 3C. The Mk 3B is for use specifically between latitudes 25° and 45°, north or south. The Mk 3Λ recorder is for use generally in latitudes within 40° of the equator.

10.1.1. Sunshine recorder Mk 2 (Plate XXIX) has a main base on which the instrument is mounted. The base has three lugs which are drilled to take screws for fixing purposes. Carried on the main base is a sub-base, consisting of a triangular plate, which in turn carries the bowl mounting and the seating for the glass sphere. The sub-base is supported on three adjusting screws so that it can be levelled accurately, and provision is made for it to be moved relative to the main base over a small range of azimuth. The solid glass sphere is seated on a cup on the top of a short pillar which is adjustable over small ranges of level and azimuth, permitting the glass sphere and bowl to be concentrically positioned. The card is held in one of three sets of overlapping grooves, one set appropriate to winter, another to summer, and a third to the equinoctial periods. The sunshine cards are held firmly in position by means of a clamping screw which is attached to a chain to prevent loss. The bowl is supported on a bracket which is slotted to provide an adjustment for latitude. A sphere-retaining clip can be fitted to prevent the displacement of the sphere by high winds or by other means, e.g. by large birds.

10.1.2. Sunshine recorders Mk 3A, 3B and 3C each consist of a solid glass sphere mounted concentrically within a portion of a spherical bowl (Mk 3C is illustrated in Plate XXX). The sphere support is a semicircular brass bar of nearly rectangular cross-section, attached symmetrically to the back of the bowl and concentric with it. The sphere is held by means of two brass screws, one fitted into a cup-shaped boss and the other fitted into a ball-ended boss, diametrically opposite on the sphere. Three overlapping pairs of grooves in the bowl accept the three types of sunshine card, the cards being secured by a clamping screw. The arc sphere support is mounted in a grooved slide to permit adjustment for latitude, and the slide is mounted on a T-shaped metal base which is supported for levelling on a fixed metal sub-base. The sub-base has three lugs which are drilled to take screws for fixing purposes.

10.2. SUNSHINE CARDS

There are three types of card; the type to be used varies with the season of the year, but otherwise any card is appropriate to any latitude and to any Mark of recorder. Figure 16 shows a cross-section of the bowl, and indicates the grooves to be used at a particular time of the year.

In the northern hemisphere the cards are used as follows:

(*a*) Long curved cards during the summer, from 12 April to 2 September inclusive; the card is inserted, with its convex side uppermost, beneath the flanges which are marked 'summer card' in Figure 16.
Note. At latitudes of 25° or less, the long curved summer cards are trimmed at each end (as instructed on the back of card) before insertion into the recorder.

(*b*) Short curved cards during the winter, from 15 October to the last day of February inclusive; the card is inserted, with its concave edge uppermost, beneath the flanges which are marked 'winter card' in Figure 16.
Note. At latitudes greater than 40°, the short curved winter cards are trimmed at each end (as instructed on the back of the card) before insertion into the recorder.

(*c*) Straight cards about the times of the equinoxes, from 1 March to 11 April inclusive, and again from 3 September to 14 October inclusive; the card is inserted beneath the central pair of flanges which are marked 'equinoctial card' in Figure 16. When inserting the equinoctial card, care must be taken to see that the hour figures are the right way up (otherwise the morning sunshine will be recorded on the portion of the card intended to receive the afternoon record, and vice versa).

In the southern hemisphere the short curved cards are for use in the winter period, 12 April to 2 September, and the long curved cards during the summer period, 15 October to the last day of February.

Before bringing a new type of card into use it is advisable to clean away any dirt which may have accumulated in the grooves into which the card will be placed.

10.3. CHANGING THE CARD

A new card is inserted into the recorder bowl each day, even when there has been no sunshine during the past 24 hours. If, after rain, the old card cannot be withdrawn without tearing it, it should be cut out carefully by drawing a sharp knife along the edge of one of the flanges.

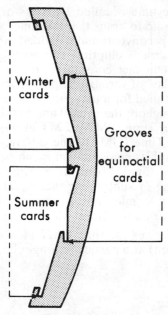

Figure 16.   Cross-section of a sunshine bowl

Ideally, the card is changed between sunset and sunrise; at stations where this done the following details are entered on the back of the card before insertion: the station name, year, month, and date of the following day. No other entries are required.

At many climatological stations the cards are changed during the day, either at 0900 GMT or (at Health Resort stations) at 6 p.m. clock time. Practices at these stations are:

(*a*) If the sun is shining at changing time, make a pencil mark from the sun's image to the lower edge of the old card before removing it.

(*b*) Slacken the clamping screw, remove the card by a sliding action along the grooves, and enter the date and time OFF on the back.

(*c*) On the back of the new card enter the station name, year, month, date and time ON. Insert the new card by sliding it along the appropriate grooves and adjust the 12 time line to coincide with the noon line engraved on the bowl, then tighten the clamping screw.

(*d*) If the sun is shining, make a pencil mark from the sun's image to the upper edge of the new card.

The pencil marks on the old and the new cards are to help in allotting the two parts of the burn to the appropriate day, and to identify any overlap of the burns.

10.3.1. Local Mean and Local Apparent Time.   It should be noted that the time indicated by the image of the sun on the sunshine card does not, in general, coincide with an hour line printed on the card. This apparent discrepancy arises for the following reasons. When the centre of the sun is due south of an observer the time is called 12 hours or noon, Local Apparent Time (LAT). The sun is said to 'transit' at this time, and the interval between two consecutive transits is conventionally divided into 24 equal parts. However, the orbit of the earth is elliptical and so the time interval between successive transits is not in fact constant throughout the year. Since it is inconvenient to have a 'day' of variable length for most practical purposes, a constant day has been defined for a mean sun whose apparent motion round the earth is uniform throughout the year. Time in this day of constant length is designated Local Mean Time (LMT); LMT at the Greenwich meridian is called Greenwich Mean Time (GMT). The difference in time between the earth's true behaviour and its conventional mean behaviour is expressed in terms of the 'equation of time'. Local Mean Time is obtained from Local Apparent Time by adding or subtracting the 'equation of time'; this varies from about +15 minutes in mid-February to about −17 minutes in mid-November for the observer on the Greenwich meridian.

Table III on page 208 gives the time GMT at which the sun will be due south (local apparent noon) at a particular longitude, east or west of Greenwich, on a particular date.

10.4. MEASURING THE TRACE

The correct method of measuring the daily duration of sunshine recorded on a card is to place the edge of a reversed card of the same type along the length of the burn and to mark on it lengths equal to those of the successive burns. The card is moved along the trace so that the lengths are added automatically, the line marking the end of one burn being used to mark the beginning of the next. The total duration is then read off on the time scale of the measured card (special care being taken to make the measurement at the same level as the burn). When the burn is not parallel to the edges of the card, because of a faulty adjustment of the recorder, the measurement should be made at the level of the trace at noon.

10.4.1. Days of broken sunshine. On a day of broken sunshine the measurement is complicated because of the spread of the burn. When the sun is shining brightly the scorch is not a fine line but a broad burn up to 3 mm wide. This spreading exaggerates the duration of short bursts of sunshine and some allowance must be made for this when measuring the record. The separate burns have rounded ends and the standard practice is to measure from a point about half-way between the centre of curvature and the extreme visible limit of the burn (see Plate XXXI).

10.4.2. Small circular burns. Small circular burns may be caused by very short bursts of sunshine, sometimes lasting for as little as two or three seconds. Such burns, which on close examination show no signs of elongation, should be ignored unless they constitute the only record on an otherwise blank card, in which case a daily total of 0·1 hour should be recorded. Thus a tabulated daily total of 0·1 hour means that the day was not entirely sunless, rather than that the duration was one-tenth of an hour.

It is sometimes difficult to decide whether or not a small burn is truly circular. In most cases when the card is burnt right through it will be found, on closer inspection, that the burn is in fact slightly elongated. When a burn appears to be made up of a series of small circular burns joined together, it should nevertheless be measured as a continuous burn, provided of course that no uncharred card is included in the measurement.

10.4.3. Faint burns.　On a day of uninterrupted sunshine the trace is a continuous burn, tapering off at the ends. This tapering off occurs because there is only a slow increase in the burning power of the sun's rays after sunrise, with a corresponding decrease before sunset. In these circumstances the measurement should be taken to the extreme ends of the tapering burn; the whole of the brown trace should be measured as far as it can fairly be seen. It should be noted that a slight discoloration of the card to light blue does not indicate bright sunshine and should therefore be ignored.

On occasions when the intensity of sunshine is diminished by haze, thin cloud or mist, periods of sunshine may be recorded as faint brown scorches similar to those seen at the ends of a record near sunrise and sunset. Because there is no spread of the burn on these occasions, the whole of each brown trace should be measured as far as it can be seen.

Under conditions of continuous weak sunshine it is possible for apparent breaks in the recorded trace to occur when the burn crosses an hour line printed on the card: these gaps should be ignored.

10.4.4. Hourly amounts.　Certain stations are required to measure hourly amounts of sunshine and to tabulate them on Metform 3445. The rules in previous paragraphs should be followed, subject to the following additions. Measurements should be made in tenths of an hour and entered on the form in accordance with the instructions printed on it.

The daily totals should be obtained by the method described at the beginning of this section (page 156), not by adding up the hourly amounts. If necessary, the hourly amounts should then be adjusted so that their sum agrees with the daily total, but in carrying out any adjustments the following restrictions should be observed.

(*a*) Completely sunless hours and hours of unbroken sunshine should be left unadjusted.

(*b*) Hours with breaks amounting to less than one-tenth of an hour should NOT be rounded up to a whole hour: they should be tabulated as ·9.

Any adjustment should then be distributed among the remaining hours having incomplete burns, in approximate proportion to their sunshine duration.

Small circular burns are ignored unless they are the only record on an otherwise blank card. In such cases the daily total is recorded as 0·1 hour, and the entry is credited to the hour in which the greatest number of circular burns occurred. If this greatest number occurred in more than one hour, the entry is credited to the hour with the most prominent burn.

10.4.5. Checking the measurements.　Consistency checks are made on all sunshine data returned to the Meteorological Office. When problems arise from either overestimates or underestimates, guidance on the correct procedures to be followed is offered to the observer concerned.

10.5. WRITING UP THE CARDS

The name of the station and the date (day, month, year) will already have been written on the back of the card. At stations where the card is changed after sunset the measurement of duration is entered in either of the two spaces provided.

10.5.1. At stations where the card is changed between sunrise and sunset, usually at 0900 GMT, the following procedure should be followed:

(*a*) Measure the amount of sunshine recorded between sunrise and time off; (the time off should have been marked with a pencil line below the trace if the sun was shining at the time). Turn the card over and enter this amount on the back in the left-hand space provided.

(*b*) Measure the duration from time on (pencil mark above the trace) to sunset. Reverse the card again and enter this duration in the right-hand space provided.

After measurement, each card will bear two amounts, each with the date to which they refer. Daily totals for entry on return forms are then obtained by adding the parts from two cards appropriate to a particular date.

10.5.2. At stations where the card is changed after sunset, but measurements of sunshine are required from sunrise to 6 p.m. clock time, and from 6 p.m. till sunset, the following routine is carried out.

When the sun is shining at 6 p.m. the position of the image of the sun is marked on the card by means of a short pencil line so that portions of the trace before and after 6 p.m. are clearly differentiated. (Since the time indicated by the recorder is Local Apparent Time, this line will not usually coincide with the printed hour line, 18 or 17, on the card.) If the two amounts are entered separately on the reverse of the card, only one date should be entered: the date on which the sunshine occurred.

10.6. CARE OF THE RECORDER

Once the instrument has been installed and properly adjusted it requires little attention beyond changing cards each day. The glass sphere should be cleaned with a soft lint-free cloth whenever necessary, and not with any material likely to abrade the surface. Excessive vigour in polishing should be avoided; impairment of the surface due to excessive polishing may result in a substantial loss of record in weak sunshine. The sphere of the Mk 2 recorder can easily be removed for cleaning. If the sphere of a Mk 3 recorder has to be removed, care must be taken to unscrew only the female (socketed) screw; if both screws are loosened the concentricity of the sphere and bowl may be lost.

The recorder should be examined each morning and any deposit, such as snow, frost, dew, bird-droppings, should be removed immediately. Hard deposits should never be chipped off since the sphere may become scratched or

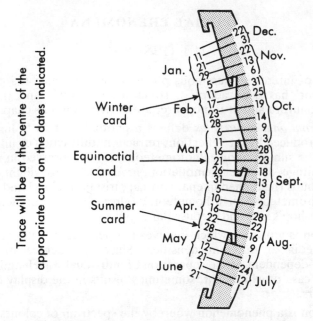

Figure 17.   Diagram showing position of
sunshine trace at different times of the year when
the latitude setting of the instrument is correct

pitted. An aerosol of windscreen de-icing fluid is recommended for defrosting the sphere which is then wiped clean. The need for adjustment may arise owing to the settlement or warping of the support and will be necessary if

(*a*) the trace is not parallel to the centre line of the card, or

(*b*) the trace is either too high or too low on the card when compared with the position of the trace on Figure 17.

Fault (*a*) is most likely to be due to an error in the level in the east–west direction. Check this first and correct the level as necessary by means of the front levelling screws. If this level is, however, found to be correct, then the fault must be due to an error in the meridian adjustment.

Fault (*b*) can be caused by incorrect latitude setting or by the recorder being out of level in the north–south plane. If the latitude setting is correct for the station, check that the sub-base is level in the north–south plane with the aid of a spirit level (a machined flat is provided on the sub-base for this purpose) and adjust by means of the rear levelling screw only. If any adjustment to either the latitude setting or the north–south level is made, it is essential to check the east–west level again.

The adjustments necessary to correct either or both of these faults are explained in I.10.3 of Appendix I (page 195). Observers are warned particularly against attempting to correct a faulty record on the card by altering the centralizing of the sphere. If an error of centralizing is suspected, the matter should be reported to the Meteorological Office which, if necessary, will make arrangements for the adjustment to be checked and corrected.

CHAPTER 11

SPECIAL PHENOMENA

11.1. GENERAL

A number of interesting luminous phenomena, for which the generic term 'photometeor' has been adopted, some special clouds and the two electrometeors, aurora and St Elmo's fire, are described in this chapter.

The *International cloud atlas* defines a photometeor as 'a luminous phenomenon produced by the reflection, refraction, diffraction or interference of light from the sun or moon'. Photometeors may be observed in more or less clear air (mirage, shimmer, scintillation, green flash and twilight colours), on or inside clouds (halo phenomena, coronae, irisation, glory) and on or inside certain hydrometeors* or lithometeors* (glory, halo phenomena, coronae, rainbow, Bishop's ring and crepuscular rays).

Refraction is a process which involves the bending of light when it passes through media of different densities. Since the degree of bending is wavelength-dependent, some separation of individual wavelengths will occur and, in the case of white light, sometimes results in the display of its constituent colours.

Diffraction is a phenomenon whereby the spectrum of colours is observed in the geometrical shadow of an obstacle. It is caused by the differential bending of the constituent wavelengths of white light which results in the spectrum. It is explained by the wave theory of light and not by the simple assumption that light travels in straight lines. Diffraction gives rise to several phenomena (see 11.2.8 to 11.2.11) arising from either a suspension of dust or water drops in the atmosphere.

Some of the phenomena described, such as the rainbow and the halo of 22°, are quite common; others are rare. An observer who is fortunate enough to see phenomena of the latter class is encouraged to make as accurate and as detailed a record as possible, including any measurements, sketches or photographs (see 11.6). Any phenomenon thought to be of exceptional interest should be reported to the Meteorological Office (Met O 15).

11.2. PHOTOMETEORS

11.2.1. Halo phenomena: a group of optical phenomena in the form of rings, arcs, pillars or bright spots produced by refraction and reflection of light by ice crystals suspended in the atmosphere (cirriform clouds, diamond dust etc.).

These phenomena, when formed by the refraction of light from the sun, may show colours, but halo phenomena produced by the light of the moon are

---

*Hydrometeors and lithometeors are defined in 4.2.1 and 4.2.2, pages 58 and 63.

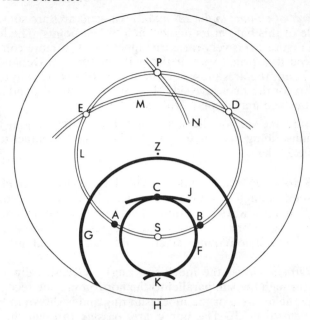

Figure 18.   Halo phenomena

An unusually complete halo display seen about midday on 6 March 1941 in the west Midlands; composite diagram based on observations from various localities.

Appearances due to refraction, which may be brilliantly coloured, as on this occasion, are shown black; appearances due to reflection, which are always white, are shown by the finer lines. The outer circle, H, is the complete horizon, with Z, the zenith, in the centre; S is the sun, P the anthelion, A, B, C, D, E are parhelia or mock suns; F is the 22° halo, G the 46° halo, J and K are upper and lower arcs of contact of F; L is the parhelic circle or mock-sun ring (parallel to the horizon), M the arc through the mock suns at 120° (usually a pair of arcs not joined in the middle), and N the oblique arc through P (one of a symmetrical pair which may sometimes be seen together).

always white. The many different kinds of halo which have been observed are shown in Figure 18 which is a composite diagram made up from a number of drawings of an unusually complete halo display seen about midday on 6 March 1941 in various localities in the west Midlands.

11.2.1.1. *Halo of 22°* (small halo) is the most frequent halo phenomenon and appears as a luminous ring, F in Figure 18, with the sun or moon, S, at its centre and having a radius of 22° of a great circle (90° is the measurement of the arc extending from the zenith to the horizon). The space within the ring appears less bright than that just outside. The ring, if faint, is white; when more strongly developed it shows coloration: the edge nearest the sun is red and this is followed by yellow and, in some rare cases, a green or violet fringe can be detected on the outside.

The angle of 22° is the angle of minimum deviation for light passing through a prism of ice with faces inclined at 60°, and this halo is probably due to the refraction of light through hexagonal prisms among ice crystals in cloud.

11.2.1.2. *Arcs of contact to the 22° halo.*   Tangent arcs are sometimes seen on the outside of this halo at its highest and lowest points. The lower arc of contact (K in Figure 18) is very rare; the upper arc, J, is more common. Both may be reduced to a bright spot at the point of contact. Depending on the altitude of the sun, the arcs present either their convex sides to the sun (low solar elevations) or their concave side (high solar elevations) and are generally very luminous and may display brilliant colour effects.

It is very rare indeed for the upper and lower arcs to join to form an elliptical circumscribing halo or for two further arcs of contact to appear at the sides of the 22° halo.

11.2.1.3. *Halo of 46°* (large halo).   This large halo (G in Figure 18) is occasionally seen, though it is seldom complete. It is much less common than the halo of 22° and is always less bright. This halo also has arcs of contact; in fact these arcs occur more frequently than the halo itself and are sometimes mistaken for it. This halo requires crystals with faces at right angles.

11.2.1.4. *Parhelic circle* (the mock-sun ring).   Occasionally a white ring which passes through the sun parallel to the horizon may be recognized. This is called the parhelic circle or the mock-sun ring and is shown in Figure 18 in its rare complete form, L. The minor arc, passing through the sun to the points of intersection with the 22° halo, may at times be distinct, faint or invisible when the major arc, or parts of it, can be clearly seen. Bright spots may be observed at certain points of the parhelic circle. These bright spots occur most commonly a little outside the halo of 22° (parhelia, at A and B), occasionally at an azimuthal distance of 120° from the sun (paranthelia, at E and D) and, very rarely, opposite the sun (anthelion, at P). When they are particularly bright they are called mock suns.

The corresponding phenomena produced by the moon are called paraselenic circle, paraselenae, parantiselenae, and antiselene; when they are bright they are called mock moons.

Mock suns shown at A and B in Figure 18 are very luminous and brilliantly coloured. Red is on the side nearest to the sun, with yellow, green, blue and violet following; the blue is generally indistinct and the violet usually too faint to be distinguished. These mock suns are situated approximately on the 22° halo when the sun's altitude is 10° or less; with increasing altitude they are formed further outside the halo, being 14° outside when the sun's altitude is 55°. They cannot be formed when the sun's altitude exceeds 60° 45'. They lie on the parhelic circle which may or may not be visible at the time. Mock suns at E and D are white. The image of the sun occasionally observed at P is a brilliant white and is sometimes termed the 'counter-sun'.

Through the mock suns D and E, two separate arms, the paranthelic arcs, may rarely be seen and they are sometimes (though very rarely) observed to join and appear to form the continuous arc M. The oblique arc N, through the anthelion P, is one of a pair; on some occasions one may be seen clearly while the other is invisible. These arcs, being caused by reflection, are white.

11.2.1.5. *Halo of 90°.*   A fourth ring, the halo of 90°, is exceedingly rare; it is not shown in Figure 18. The ring is white and cannot be seen in its entirety unless the sun is at the zenith.

11.2.1.6. *Circumzenithal arc.*   Occasionally the upper and lower circumzenithal arcs may be observed; they appear to lie in horizontal planes. The upper circumzenithal arc (brightly coloured, with red on the outside and violet on the inside) is a rather sharply curved arc of a small horizontal circle near the zenith; the lower circumzenithal arc is a flat arc of a large horizontal circle near the horizon. The upper arc occurs only when the angular altitude of the luminary is less than 32°; the lower arc occurs only when the angular altitude of the luminary is more than 58°. The upper arc touches the large halo, if visible, when the angular altitude of the luminary is about 22°; the lower arc touches the large halo when the angular altitude of the luminary is about 68°. The arcs become increasingly separated from the large halo as the angular altitude of the luminary departs from the above values. Circumzenithal arcs may be observed without the large halo being visible.

11.2.1.7. *Sun pillar.*   This may be seen occasionally, particularly at sunrise or sunset, and will frequently extend about 20° above the sun and generally ends in a point. At sunset it may be entirely red, but is usually a blinding white and shows a marked glittering. If the sun is high in the heavens the pillar may appear as a white band vertically above or below the sun, but it will not be very brilliant and is often short. Occasionally, however, these white columns appear simultaneously with a portion of the white mock-sun ring, and so form another remarkable phenomenon, the cross. Sun pillars are due to reflection of sunlight from ice crystals.

11.2.1.8. *The undersun.*   This is a halo phenomenon produced by reflection of sunlight on ice crystals in cloud. It appears vertically below the sun in the form of a brilliant white spot, similar to the image of the sun on a calm water surface. It is necessary to look downward to see the undersun; the phenomenon is therefore only observed from aircraft or from mountains.

11.2.2. Rainbows: a group of concentric arcs with colours ranging from violet to red, produced on a 'screen' of water drops (raindrops, droplets of drizzle or fog) in the atmosphere by light from the sun or moon.

This phenomenon is mainly due to refraction and reflection of light. When rainbows are produced by the sun their colours are usually brilliant; when produced by the moon their colours are much weaker or sometimes absent.

A primary rainbow is a coloured bow which appears on a 'screen' of water drops when light from the luminary falls upon them. The coloured bow is opposite the luminary by which it is produced and its centre is on the extension of the line joining the luminary and the observer. Thus the rainbow may form a complete ring when seen from an aircraft. Close inside the primary bow there may appear one, two or more supernumerary bows with the colours in the same order but with each inner bow becoming progressively much fainter. These supernumerary bows, usually narrow with green, violet or orange colours, are due to interference, and on rare occasions may also occur outside the secondary bow (see below).

It is rarely that all the 'colours of the rainbow' (red, orange, yellow, green, blue, indigo and violet) are observed. The size of the drops or droplets determines which colours are present and the width of the band occupied by each of them. Drops larger than 1 mm in diameter yield brilliant bows with violet on the inside (radius of the arc 40°) and red on the outside (radius of the arc 42°). With drops about 0·3 mm in diameter the outside limiting colour is orange, and inside the violet there are bands in which pink predominates. With smaller drops supernumerary bows appear to be separate from the primary bow. With still smaller drops about 0·05 mm in diameter the rainbow degenerates into a white fogbow with faint traces of colour at the edges. The variations of colour with drop size are due to diffraction. No primary bow can form if the altitude of the luminary is higher than 42°.

Outside the primary bow there may be a secondary bow, much less bright than the primary and with a breadth about twice that of the primary. The red is on the inside (radius of the arc 50°) and the violet is on the outside (radius of the arc 54°). No secondary bow can form if the altitude of the luminary exceeds 54°.

Reflection rainbows are occasionally seen on calm days when a sheet of water lies in front of or behind the observer when standing with his back to the sun. Such bows are formed by rays of light illuminating the falling raindrops after reflection at the surface of a sheet of water. The centre of a reflection above is thus as high above the horizon as the sun, or the same angular distance above the horizon as the centre of the direct bow is below the horizon; consequently the arc, when complete, exceeds a semicircle. The direct and reflection rainbows intersect on the horizon, and the colours have the same sequence.

11.2.3. Mirage: an optical phenomenon consisting mainly of steady or wavering, single or multiple, upright or inverted, vertically enlarged or reduced, images of distant objects.

Mirages are produced by refraction of light in the layers of air close to the earth's surface due to large temperature gradients in the vertical and associated changes of refractive index. Objects seen in a mirage sometimes appear appreciably higher or lower above the horizon than they really are; the difference may amount to as much as 10°. Objects located below the horizon or hidden by mountains may become visible ('looming'); objects which are visible under normal circumstances may disappear during the occurrence of a mirage.

An inferior mirage (or lower mirage) may occur over a flat strongly heated surface, for example a desert, and may often be seen over road surfaces heated by the sun. Rays of light from distant objects, impinging on the heated surface at grazing incidence, are bent upwards in traversing the strongly heated layer of air on the surface, just as if they were reflected by a mirror lying horizontally on the surface. An inverted image of the object is therefore seen and the illusion of a sheet of water is created. The images are often elongated vertically or otherwise distorted.

A superior mirage (or upper mirage), in which the light rays are bent downwards from a warm stratum of air resting on a colder one, is seen most frequently in polar regions. The distances involved are greater, so that the details can hardly be observed without telescopic aid. With such assistance a distant ship, for instance, may sometimes be seen in triplicate, one of the images being inverted. As the stratification which produces a superior image is stable, the images are clear and well defined in contrast to the shimmering images of the inferior image.

The images produced by a superior mirage are not always exactly in the same vertical. The separation is explained by a slight departure of the strata of equal density from horizontal planes. Lateral refraction of this kind is to be distinguished from the phenomenon which can be observed sometimes near a heated wall when objects appear to be reflected in the wall.

Fata Morgana is a complicated form of mirage in which multiple images of an object are produced by several layers of air of different refractive indices.

11.2.4. Shimmer: the apparent fluttering of objects at the earth's surface, when viewed in the horizontal direction.

Shimmer occurs chiefly over land when the sun is shining brightly. It is due to short-period fluctuations of the refractive index in the surface layers of the atmosphere. Shimmer may reduce the visibility appreciably.

11.2.5. Scintillation: rapid variations, often in the form of pulsations, of the light from stars or terrestrial light sources. Scintillation, or twinkling, is the more or less rapid change of apparent brightness of a star, sometimes accompanied by colour changes at altitudes less than about 50°. It is due to minor variations in the refractive index of the atmosphere, mainly in the low atmosphere. The amount of twinkling is always greatest towards the horizon and least at the zenith.

Scintillation is also observed in terrestrial lights. Shimmer seen near the ground on a hot day is akin to it.

The bright planets do not usually appear to twinkle as they have discs of definite size. Each point on the disc twinkles independently of the others, so that on the average the light is steady.

11.2.6. Twilight colours: various colorations of the sky and of the peaks of mountains at sunrise and sunset.

Twilight colours are produced by refraction, scattering or selective absorption of light rays from the sun in the atmosphere.

When clouds, particularly high-level and middle-level clouds, occur about the time of sunset or sunrise, or in bright twilight, their coloration is often very beautiful, The cloud colours are mainly shades of orange, rose or red, since the direct sunlight illuminating the cloud has passed through a great length of the lower layers of the atmosphere. Shades of purple are sometimes seen, since a cloud may at the same time be indirectly illuminated by scattered blue light from higher atmospheric levels.

Colour phenomena also occur in a cloudless sky during the twilight periods. These vary considerably and are best developed in arid or semi-arid land regions. Some of those which occur most commonly everywhere are mentioned here. The *primary twilight arch* appears, after the sun has set, as a bright but not very sharply defined segment of reddish or yellowish light resting on the western horizon. After the sun has set, a pink or purple glow may be seen covering a considerable part of the western sky; this is the *first purple light.* It reaches its greatest extent and luminance when the sun is about 4° below the horizon, and disappears when it is about 6° below.

At sunset, a steely-blue segment, darker than the rest of the sky, begins to rise from the eastern horizon. This is the shadow of the earth thrown by the sun on to the earth's atmosphere. The *earth shadow* is bordered by a narrow band of rose or purple colour, called the *counterglow*. The whole rises fairly quickly in altitude, the shadow encroaching on the counterglow and soon obliterating it. With increasing general darkness the edge of the shadow weakens, but may sometimes be traced up to its passage through the zenith. In the later stages of twilight, this shadow edge comes down nearly to the westerly horizon, leaving a slightly more luminous segment between it and the horizon; this is the *secondary twilight arch*. Just before the ending of astronomical twilight (which ends when the sun's centre is 18° below the horizon), it is sometimes seen as a fairly well-defined whitish arch on the horizon, with an altitude of only a few degrees at its apex. This might be confused with an auroral arc visible at very low altitude.

Other colours are often seen in the cloudless twilight sky, portions of which may be green, yellow, orange, or red, according to the amount of dust and water vapour present in the air. Instead of the purple light after sunset, the sky very often shows some shade of clear green, probably when the air is relatively free from dust.

Analogous phenomena, in the reverse order, occur before sunrise.

11.2.7. Alpine glow: a series of phenomena seen in mountainous regions about sunrise and sunset. Two principal phases are generally recognized:

(i) The true alpine glow. At sunset the phase begins when the sun is 2° above the horizon; snow-covered mountains in the east are seen to assume a series of tints from yellow to pink, and finally purple. As this phase is due mostly to direct illumination by the sun, it terminates when the mountain tops pass into the shadow of the earth. The alpine glow is most striking when there are clouds in the western sky and the illumination of the mountains is intermittent.

(ii) The afterglow. This begins when the sun is 3° or 4° below the horizon. The lighting is faint and diffuse with no sharp boundary and occurs only when the 'purple light' is manifest in the sky.

11.2.8 Green flash: a predominantly green coloration of short duration, often in the form of a flash, seen at the extreme upper edge of a luminary (sun, moon, or sometimes even a planet) as it disappears below or appears above the horizon.

Flashes up to an altitude of several degrees have sometimes been observed. Although the colour of the phenomenon is predominantly green, blue and violet may also be visible, particularly when the air is very transparent. The colour lasts two or three seconds at the longest.

No completely satisfactory explanation has so far been given for the green flash but it is most probable that the greater refraction of the short waves (violet, blue, green) than of the long waves (red) of white sunlight, coupled with the greater degree of Rayleigh scattering experienced by the violet and blue rays, plays the most important part. It is possible that the flash may appear blue or violet in a hazy atmosphere when such differential scattering may not be appreciable. It is probable that an unusual degree of refraction, such as occurs with a low-level inversion of temperature, is required for the phenomenon.

Differential refraction of white light is also the cause of the analogous very rare 'red flash' which may occur when the sun's disc appears just below a bank of clouds near the horizon.

11.2.9. Coronae: one or more sequences (seldom more than three) of coloured rings of relatively small diameter, centred on the sun or moon. In each sequence the inside ring is violet or blue and the outside ring is red; other colours may occur in between. The innermost sequence usually shows a distinct outer ring of reddish or chestnut colour called the 'aureole', the radius of which is generally not more than 5°. Often the aureole alone appears.

Coronae are due to the diffraction of light by very small water drops. The smaller the drops the greater the radii of the aureole and of the successive, approximately equidistant, red rings. The more uniform the size of the drops the purer are the colours of the coronae.

The diameters of coronae are generally considerably smaller than that of the halo of 22° so that they are very near the luminary and can be seen around the sun only when its brightness is very much dimmed. Coronae are seen most often round the moon though no doubt they occur as frequently round the sun. Sometimes they may be more favourably observed round the sun by using a reflector (such as a pool of water or black glass) to reduce the intesity of the light.

Coronae may be distinguished from haloes by the reversal of colour sequence, the red of the halo being inside, that of the corona outside. Size of the rings should not be used as a criterion because the diameter of a corona may occasionally be quite as large as that of a halo. As coronae are produced by diffraction they occasionally show the sequence of colours two, three, or even four times over. This cannot happen with haloes.

11.2.10. Bishop's ring: a whitish ring, centred on the sun or moon, with a slightly bluish tinge on the inside and reddish-brown on the outside.

In the middle of the day the inner radius of the ring is about 10°, the outer 20°; when the sun is low the ring is larger. Bishop's ring was first seen after the great eruption of Krakatoa in 1883 and could be seen until the spring of 1886. It was also seen after the eruptions of Soufrière in St Vincent and Mt Pelée in Martinique in 1902, after the north Siberian meteorite in 1908, and at the time of nearest approach to the earth of Halley's Comet on 18 and 19 May 1910. The phenomenon is attributed to diffraction associated with fine dust in the high atmosphere. The colours of a Bishop's ring are not very distinct; they are particularly faint in rings observed around the moon, showing only a pale red fringe.

11.2.11. Glory: one or more sequences of coloured rings seen by an observer around his own shadow on a cloud consisting mainly of numerous small water drops, on fog or, very rarely, on dew.

It happens frequently on mountains that there is a mist on one side of a ridge and not on the other. In such circumstances an observer standing with his back to the sun will sometimes see coloured rings of light round the shadow of his own head on the mist. The coloured rings, called 'glory', have no direct connection with the shadow, but are caused by light diffracted backwards in the same way as the coronae are caused by light diffracted forwards. The colour order is the same as that for a corona.

When the shadow seems to be very large, because the cloud or fog are near the observer, it is called a 'Brocken spectre' whether a coloured glory is seen or not. A large outer ring, known as 'Ulloa's ring', which is essentially a white rainbow, is sometimes seen at the same time.

Airborne observers sometimes see a glory around the shadow of the aircraft in which they are flying.

11.2.12. Irisation (iridescence): colours appearing on clouds, sometimes mingled, sometimes in the form of bands nearly parallel to the margin of the clouds; green and pink predominate, often with pastel shades.

Irisation colours, often brilliant, resemble those observed on mother-of-pearl. Within about 10° from the sun, diffraction is the main cause of irisation. Beyond about 10°, however, interference is usually the predominant factor. Irisation extends at times to angles exceeding 40° from the sun and even at this angular distance the colours may be brilliant.

Irisation is occasionally seen on the edges of some species of cirrocumulus, altocumulus, and stratocumulus clouds. The coloured patches may be parts of coronae probably due to the local occurrence of exceedingly small water droplets. The point to note is the angular distance from the sun or moon of the patches which show irisation. This may be judged by holding a centimetre rule at arm's length when each centimetre will approximate to 1°.

11.3. SPECIAL CLOUDS

Nacreous and noctilucent clouds occur at high altitudes and may be observed from comparatively restricted geographical locations. Both may be seen from the British Isles but usually only in the higher latitudes.

11.3.1. Nacreous clouds: clouds resembling cirrus or altocumulus lenticularis and showing very marked irisation, similar to that of mother-of-pearl; the most brilliant colours are observed when the sun is several degrees below the horizon.

Nacreous clouds are a rare type of stratospheric cloud and were previously termed 'mother-of-pearl clouds'. The nature of the cloud particles is not yet known, but the associated optical effects suggest diffraction by spherical particles of diameter less than 2·5 micrometres; it has been suggested that the particles may be water droplets or spherical ice particles.

These clouds have been observed mainly in Scotland and Scandinavia but have occasionally been reported from Alaska and from France during periods of a strong west to north-west airstream. Measurements by Professor Störmer in southern Norway and Dr D. H. McIntosh in Scotland indicate that the clouds occur at an altitude between 21 and 30 km and require a temperature approaching −90 °C to form. Such a temperature is only likely to be approached occasionally during December, January and, less likely, February in the northern hemisphere.

The clouds are somewhat lenticular in form, very delicate in structure, and show brilliant irisation at angular distances up to about 40° from the sun. Irisation reaches its maximum brilliance when the sun is several degrees below the horizon. Later, with the sun still further down, the various colours are replaced by a general coloration which changes from orange to pink and contrasts vividly with the darkening sky. Still later, the clouds become greyish, then the colours reappear, though with greatly reduced intensity, before they finally fade out. Up to about two hours after sunset, the nacreous clouds can still be distinguished as tenuous grey clouds standing out against the starry sky. If there is moonlight, they may be visible throughout the night. Before dawn, this sequence is repeated in reverse order.

By day, nacreous clouds often resemble pale cirrus. If, after sunset, cirrus and nacreous clouds coexist, the nacreous clouds still show bright colours after the cirrus has already turned grey. Nacreous clouds show little or no movement; this fact, together with the circumstances of their occurrence strongly suggests that they are in the nature of mountain wave clouds.

11.3.2. Noctilucent clouds: clouds resembling thin cirrus, but usually with a bluish or silvery colour, or sometimes orange to red; they stand out against the night sky. Most of the information in this section is based on a review by the late J. Paton.

These clouds occur at great heights, about 80 km, and they remain sunlit long after sunset. They are visible only at night in that part of the sky where they are directly illuminated by sunlight and when the sky background is dark enough to permit their weak luminescence to be seen.

Noctilucent clouds have been observed only in latitudes higher than 45°N. The northern limit is uncertain because of prevailing cloudiness and the lack of observing sites but is probably at least 80°N. At about latitude 55°N the clouds are seen most frequently in late June and early July. They are usually observed only in the summer months.

Because of their great height it had been assumed that they consisted of volcanic or meteoric dust but sounding rockets fired in northern Sweden during the summer of 1962 successfully collected noctilucent cloud particles which on examination provided strong indications that the clouds consisted of ice crystals.

Noctilucent clouds usually appear in the form of thin cirrus-like streaks, sometimes only one or two isolated filaments being visible while at other times the cloud elements are closely compacted into an almost continuous mass resembling cirrocumulus or altocumulus undulatus. Weaker and more tenuous displays have the form of decayed cirrus and there is often a structureless background (nebula) rather similar to cirrostratus.

The noctilucent clouds can usually be distinguished from ordinary clouds by the fact that they remain brighter than the sky background and glow with a pearly, silvery light, generally showing tinges of blue coloration. They sometimes appear reddish in the immediate vicinity of the horizon.

11.3.3. Observations of nacreous and noctilucent clouds. When nacreous or noctilucent clouds are observed, observers should record (i) the night of occurrence specified by two dates, e.g. 11–12 July 1982, and the latitude and longitude of the place of observation; (ii) the period(s) of time, GMT, during which the clouds were observed; (iii) the horizontal and vertical extent, expressed in degrees of azimuth and elevation, at specified times, say every quarter-hour, half-hour or hour, and (iv) general notes on the nature and behaviour of the clouds.

It is most useful to draw a rough sketch showing the configuration of the cloud elements and the co-ordinates, elevation and azimuth of the visible boundaries of the cloud, i.e. the maximum elevations in different azimuths and the limiting azimuths, east and west of north. Advice on using a pilot-balloon theodolite, and also the use of a centimetre scale as a rough guide, in deriving the values of elevation and azimuth is given in 11.6 on page 175.

When stars are present in the vicinity of the clouds their position in relation to the clouds should be noted, thereby providing reference marks from which the altitude of the clouds may be determined. The horizon provides another useful reference mark.

If monochrome or colour photographs can be taken, a record should be made of the time to the nearest minute of each photograph, the focal length of the lens, and the azimuth and angular elevation of any landmarks from which the direction of the optical axis may be determined. With fast monochrome film, exposure times are of the order of 10 seconds at f/3.5; with colour film of rating ASA 25, exposure times are 40–60 seconds at f/3.5. It is advisable to take several photographs at different exposures.

Meteorological Office Headquarters are always interested in receiving photographs and reports of such clouds.

11.4. ELECTROMETEORS

11.4.1. St Elmo's fire (or corposant): a more or less continuous, luminous electrical discharge of weak or moderate intensity in the atmosphere, emanating from elevated objects at the earth's surface (lightning conductors, wind vanes, masts of ships) or from aircraft in flight (wing-tips etc.).

It is also seen on projecting objects on mountains. This phenomenon may be observed when the electrical field near the surface of objects becomes strong. According to some observations, the character of St Elmo's fire varies with the polarity of the electricity being discharged. The negative 'fire' is concentrated so that a structureless glow envelops an elongated object such as a mast or an aerial; the positive 'fire' takes the form of streamers about 10 cm long. It may also appear as luminous globes, a number of which are sometimes seen along the aerial. The colour of St Elmo's fire can be violet or greenish.

The phenomenon is also termed 'corposant' (*corpo santo:* holy body) because of its once-supposed supernatural nature.

11.4.2. Aurora: a luminous phenomenon which appears in the high atmosphere in the form of arcs, bands, draperies, curtains or rays.

Aurorae are due to electrically charged particles ejected from the sun and acting on the rarified gases of the high atmosphere. As the charged particles spiral towards the earth, their visible interaction with the gases reaches its maximum frequency in the 'auroral zones'; these zones are centred on the geomagnetic poles with a radius of approximately 20° and aurora is visible near them on almost every clear night. The northern auroral zone lies just north of Norway, south of Iceland and Greenland, over northern Canada and north of Siberia.

Measurements have indicated that the altitude of the lower limit of polar aurorae is approximately 60 to 100 km and can extend to as high as 1000 km. The frequency of aurora varies markedly with solar activity, tending to occur at times of maximum sunspot number; at times of minimum sunspot number the auroral displays seem to recur at 27-day intervals, these being the periods of rotation of the sun. The extent of auroral activity is also much enhanced during a major solar flare or storm when aurora may be visible far to the south of its normal position. The estimated annual frequency of nights of visible aurora, ignoring the intervention of cloud, twilight or moonlight, is about 150 in northern Scotland and about 10 in southern England. The corresponding frequencies of overhead displays are about 10 and 1 respectively. The terms 'aurora borealis' or 'northern lights' and 'aurora australis' or 'southern lights' apply respectively to the northern and southern hemispheres. Aurora is usually pale and grey but sometimes, during more active periods, it will become yellow-green or red.

The common forms of display and their descriptions are given in the drawings of Figure 19. Aurora often appears as a *glow* (a) on the poleward horizon; such a glow is the upper part of some other auroral form whose lower edge is below the horizon. A common form, particularly in high altitudes, is an *arc;* this may be *homogeneous,* i.e. uniform in brightness (b), or *rayed,* i.e. with vertical ray structure (c). Multiple arcs, running parallel to one another across the sky, are not uncommon. When an arc has folds along its length it is called a *band;* this also may be *homogeneous* (d) or *rayed* (e). Rayed bands in which the rays are long look like curtains or draperies. Single *rays* (f) or bundles of rays are often seen after the break-up of a rayed arc or band. If the display passes overhead, the parallel rays along the lines of force appear, by perspective, to converge at the magnetic zenith, thus producing *coronal rays* (g). The magnetic zenith is not coincident with the observer's zenith but is displaced towards the equator by an amount depending on the distance of the observer from the geomagnetic pole. In the British Isles the magnetic zenith is about 20° south of the observer's zenith. During a display brighter patches may occur and move along the arc or drapery and the process of 'flaming' may be seen in which light surges upwards from the lower edge of the arc or drapery to the zenith. As a display fades the aurora will degenerate into *patches* (h) which are like diffuse clouds without any arc or ray structure.

The distribution of aurorae in time and space has been determined by visual observation and by simultaneous photography from a number of stations, some of which have been equipped with automatic 'all-sky' cameras. Radio-echo techniques have also been used in investigating aurorae.

The observer should give a description of the aurora and note the direction in which it appears most intense, the position and angular height above the

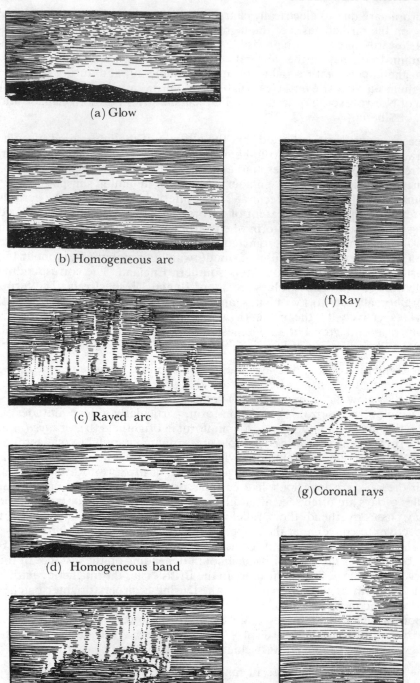

(a) Glow

(b) Homogeneous arc

(f) Ray

(c) Rayed arc

(g) Coronal rays

(d) Homogeneous band

(e) Rayed band

(h) Patch

Figure 19.   Auroral forms

horizon of particular features, and colour effects. The advice about sketches and photographs given in 11.3.3 for nacreous and noctilucent clouds also applies to aurorae.

11.4.2.1. *Dating of aurorae.* Auroral displays are dated according to the date of commencement of the night on which the display occurs. For example, an aurora beginning on the 25th of the month and ending on the 26th is described as 'the aurora of the 25th'. An aurora occurring after midnight may, however, be referred to as 'the aurora of the early hours of the 26th' in a case where ambiguity may otherwise arise. In statistical summaries, statements referring to frequency are in the form 'aurora was observed on X nights'.

11.5. MISCELLANEOUS PHENOMENA

11.5.1. Crepuscular rays. Three similar phenomena are included under the term 'crepuscular rays'. These are:

(a) Sunbeams penetrating through gaps in a layer of low cloud and rendered luminous by water or dust particles in the air (phenomenon termed 'sun drawing water' or 'Jacob's ladder').

(b) Pale blue or whitish beams diverging upwards from the sun hidden behind cumulus or cumulonimbus clouds. The well-defined beams are separated by darker streaks which are the shadows of parts of the irregular cloud.

(c) Red or rose-coloured beams, diverging upwards at twilight from the sun below the horizon. The light is scattered to the observer by atmospheric dust; the beams are separated by greenish-coloured regions which are the shadows of clouds or hills below the horizon.

In classes (b) and (c) the beams and shadows may persist across the sky before converging at the 'antisolar point' (anti-crepuscular rays). The apparent divergence and convergence is an optical illusion produced by perspective.

11.5.2. Day darkness. The term 'day darkness' or 'high fog' describes a state of gloom caused by a concentration of smoke at no great height (a few hundreds of metres or feet) above ground level. In the British Isles it is a winter phenomenon, now relatively infrequent, characteristic of large conurbations. Although the darkness may be comparable with that of night, the visibility near the ground may not be seriously impaired.

The 'special phenomenon' group for reporting day darkness (Regional Code 668 in *Handbook of weather messages*, Part II) provides for three degrees of day darkness, 'bad', 'very bad' and 'black' (Regional Code 624) which are self-explanatory, and also provides for reporting the direction in which the day darkness is worst. The phenomenon is different from the ordinary gloom of a murky day caused by dense low cloud and poor visibility. The darkening of the sky is abnormal and often shows a dirty greenish or reddish coloration even when the darkness is no worse than 'bad'. In 'black' day darkness there is no coloration and the pall appears impenetrable by the sun's rays.

11.5.3. Zodiacal light and associated phenomena. The zodiacal light is observed as the cone-shaped extremity of an elongated ellipse of soft whitish light. It extends from the sun as centre and appears above the westerly horizon after sunset or above the easterly horizon before sunrise. The light retains its apparent place among the stars, as it gradually sets in the evening or rises in the morning. It is most brilliant and most readily seen in the tropics, and it may sometimes be conspicuous in temperate latitudes provided that it is observed away from the glare of large towns. It is then often sufficiently bright to be visible to some extent in faint twilight or faint moonlight.

The light is pearly and homogeneous, and differs markedly in quality from that of the Milky Way, the brightest part of which it may considerably exceed in luminance. Its luminance decreases with altitude, since its brightness is greater the nearer the observed point is to the position of the sun below the horizon. It appears, however, to fall off in brightness very near to the horizon on account of the greater thickness of the atmosphere its light has to traverse. At any particular altitude the axis of the light is brighter than its lateral parts. The best time for observation is just after the last traces of twilight have disappeared in the evening, or just before the first traces appear in the morning, the greatest extent of the light being visible.

The axis of the light lies very nearly, but not quite, in the plane of the ecliptic, and the whole phenomenon is thus confined to the zodiacal constellation. In temperate latitudes the ecliptic is often inclined at such a small angle to the horizon that the light is rendered invisible by the additional extent of the atmosphere traversed. In the latitudes of the United Kingdom it is best seen in the evenings of January to March and in the mornings of September to November; only about the time of the winter solstice is it possible to observe it on the morning and evening of the same day. In February the light lies between the constellations of Pegasus and Cetus, the apex being near the Pleiades. Just after dark in this month, in the latitudes of the United Kingdom, the altitude of the apex of the light is about 50°, the cone lying obliquely with regard to the horizon. The breadth near the horizon at this time is 25° to 30°, and the altitude of the brightest part 20° to 30°. At this time the light has a fairly clear-cut boundary on the side towards the south; on the side towards the north it is more diffuse, and is often seen to have an extension to the north for some distance above the horizon. Observations in other latitudes, and especially in the tropics, have not been frequent enough to determine whether this difference of the two edges and the extension to the north are seen everywhere.

The zodiacal light is generally believed to be caused by the scattering of sunlight from a cloud of particles lying in the ecliptic. The composition and origin of these particles is not yet certain but they are believed to be of interplanetary dust which is the debris of comets.

11.5.3.1. *Zodiacal band and Gegenschein.* Associated with the zodiacal light are two other phenomena, the zodiacal band and the Gegenschein. Joining the apexes of the morning and evening zodiacal lights is an extremely faint luminous band, a few degrees wide, lying along or nearly along the ecliptic; this is called the zodiacal band. On it at a point nearly, or perhaps exactly, 180° from the sun's position in the ecliptic, is a somewhat brighter and larger, though rather ill-defined, patch of light, 10° or more in diameter, known by the German name Gegenschein (literally 'counterglow', but this term in English is applied to a meteorological phenomenon of twilight, see 11.2.6.). When they are at a sufficient altitude, the zodiacal band and the Gegenschein may be observed in temperate latitudes on the clearest moonless nights. The position and width of the band and the size, shape and exact position of the Gegenschein should be noted, together with variations of brilliancy and any special features seen, but the observation will be found difficult even for the keenest eyesight. The Gegenschein is usually invisible for the few nights on which it is projected upon the Milky Way in its annual journey round the ecliptic.

11.6. RECORDING PHENOMENA

For the more unusual phenomena an observer may record such information as the times of occurrence at different stages or phases, angular diameters of circles or arcs, notes on their colouring, angular elevation, movement, etc. Such information is part of the record of a station and should be retained so that it can be referred to by any interested party. Some notes are given here on the best way to measure angles both accurately and approximately. Photographs provide a suitable means of recording some phenomena and advice is included on exposure, but this must be considered as providing only a rough guide.

If a pilot-balloon theodolite is available, the elevation scale can be used to measure angles in the vertical plane. Caution must be exercised, however, in using the azimuth scale to measure the angular radii of haloes etc. If A_1 and E are the azimuth and elevation of the sun and A_2 is the azimuth of the limb of the halo at the same elevation, the angular radius (α) of the halo is not $A_2 - A_1$ but may be derived from the expression

$$\sin \alpha = \cos E \sin (A_2 - A_1).$$

Rough angular measurements may be made by holding a centimetre scale at arm's length and taking one centimetre as equivalent to one degree. The exact distance for which this is true is 57·4 cm (22·6 inches).

11.6.1. Photographing phenomena.   Advice on photographing nacreous and noctilucent clouds is given in 11.3.3.

Colour transparencies are ideal for recording clouds and many optical phenomena. It is essential that the correct light setting is used, and the photographer should not be surprised if a very high light value is measured. The light from clouds and halo phenomena is very intense, and fast shutter speeds with small apertures are frequently required even with slow colour film. When photographing clouds it is often helpful if the horizon or some foreground feature can be included to give perspective to the picture. Part of a nearby tree may be ideal. If a halo is to be photographed it is necessary to obscure the sun's disc with some intervening object unless the sun is low in the sky. A slender tree, a chimney stack or small amounts of any lower cloud of sufficient optical thickness may be utilized. The standard lens of a 35 mm camera will not photograph a complete halo on one frame.

Cloud photography on black and white film is also possible but here again the correct exposure setting must be used. In addition an orange (×5) or red (×10) filter will accentuate the blue tones when blue sky is visible and give a better picture for scientific purposes. Even a yellow (×2) filter will help in this respect. The recording of optical phenomena is more difficult with black and white film than with colour, and unless one is experienced in this work a more reliable result will be obtained by making a sketch. It must be remembered too that colours do not record well on monochrome film; for example, a bright red will appear black on the print.

Whenever a photograph is taken, the observer should record the time, place, direction the camera was pointing, aperture, shutter speed, type of film and filter (if used) together with suitable notes of the special points observed.

APPENDIX I

REQUIREMENTS AT OBSERVING STATIONS

I.1. AUTHORITY AND OBSERVER

The word 'authority' is used to denote the person, organization or corporate body responsible for the maintenance of the station or, in the case of an auxiliary station, by whose permission the station co-operates with the Meteorological Office. The 'observer' is the person to whom is assigned the actual duty of making and recording the observations. At a private station the authority and the observer may of course be one and the same person.

Where stations are maintained by organizations or corporate bodies such as Town Councils, the authority is asked to nominate a particular person with whom the Meteorological Office should correspond on matters relating to the station. It is usual for minor queries concerning the observations etc. to be sent direct to the observer.

There should also be a deputy observer capable of taking over the observer's duties in the latter's absence. It is a good plan to arrange for the deputy to take spells of duty periodically so that his knowledge of the procedure does not become rusty from disuse.

Both the observer and his deputy should have ready access to this handbook which is intended as a manual for regular use.

I.2. SELECTION OF SITE

To ensure that the observations are representative of the place and sufficiently comparable with those made at other stations to permit their use in national or regional studies, the following basic requirements are laid down for synoptic and climatological stations.

(a) Outdoor instruments should be installed on a level piece of ground, approximately 10 m by 7 m, covered with short grass, and surrounded by open fencing or palings to exclude unauthorized persons. Within this enclosure a bare patch of ground about 2 m by 2 m is reserved for observations of the state of ground and of soil temperature at depths of less than 30 cm. A suggested layout is shown in Figure 20.

(b) There should be no steeply sloping ground in the vicinity and the site should not be in a hollow. If these conditions are not complied with, the readings of temperature and amount of precipitation may show peculiarities of entirely local significance.

(c) The site should be well away from trees, buildings, walls or other obstructions. The distance of any such obstacle (including fencing) from the rain-gauge should not be less than twice the height of the object above the rim of the gauge, and preferably four times the height.

177

(*d*) The sunshine recorder, rain-gauge and anemometer must be on sites with exposures to satisfy their requirements and they need not be on the same site as the other instruments.

(*e*) As noted in 5.3 (page 84), the enclosure may not be the best place from which to estimate the wind speed and direction; another observation point, more exposed to the wind, may be desirable.

(*f*) Very open sites which are satisfactory for most instruments are unsuitable for rain-gauges. For such sites the rainfall catch is reduced in other than light winds and some degree of shelter is needed (see I.9.1 and I.9.2).

(*g*) If the instrument enclosure does not command a sufficiently extensive view over the surrounding country, alternative viewpoints should be selected for observations of visibility.

The position used for observing cloud and visibility should be as open as possible and command the widest possible view of the sky and the surrounding country.

In selecting a site the future should be considered as well as the present. A good site may become a bad one because of the growth of trees or the erection of buildings on adjacent plots. Where the station is owned by an urban authority it should preferably be sited on a scheduled open space, and layout of the remainder of the open space should be such that the exposure will remain unimpaired for many years.

At stations on aerodromes, apart from the requirements set out above, the enclosure should be in close proximity to the meteorological office for convenience of access, but it should be more than 300 m downwind (with respect to the prevailing wind) of any area used for running aircraft engines, and should be more than 100 m in any direction from any such area.

At coastal stations it is desirable that the station should command a view of the open sea, but should not be too near the edge of a cliff because the wind eddies created by the cliff will affect measurements of amount of precipitation and wind.

Night observations of cloud and visibility are best made from a site unaffected by extraneous lighting.

I.3. CO-ORDINATES OF THE STATION

For the record and other purposes the position of a station must be accurately known. The 'co-ordinates' of a station are

(*a*) the latitude to the nearest minute,

(*b*) the longitude to the nearest minute, and

(*c*) the height of the station above mean sea level, i.e. the altitude of the station, to the nearest metre.

These co-ordinates refer to the plot on which the observations are taken and may not be the same as those of the town, village or airfield after which the station is named.

For a station in Great Britain the latitude and longitude can be read from the appropriate sheet of the 1:50 000 Ordnance Survey map. The position of

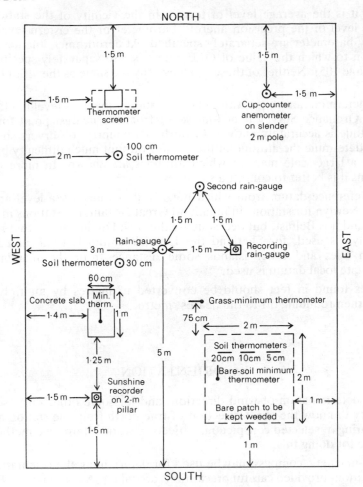

Figure 20. Layout of an observing station

the station is more closely identified by the National Grid reference for the site and this should also be given, as read from the appropriate 1:50 000 Ordnance Survey map in accordance with the instructions which are given at the side of each sheet. For a station in Northern Ireland, the latitude, longitude and the Irish Grid reference should be read from the appropriate one-inch map published by the Ordnance Survey of Northern Ireland.

For a station elsewhere, a grid reference should be given if possible, but the grid used must be specified, and the date of the map should be given to avoid confusion with alternative or obsolete grids.

The altitude of the station is defined as the height above mean sea level of the ground on which the rain-gauge stands or, if there is no rain-gauge, the ground beneath the thermometer screen. If there is neither rain-gauge nor screen, it is the average level of terrain in the vicinity of the station. The capsule level of the precision aneroid barometer or the cistern level of the mercury barometer are separately specified. At aerodromes, the aerodrome elevation to which the value of QFE refers is also separately specified (see 7.5.1, page 105). Neither of these is necessarily the same as the altitude of the station.

The determination of the altitude of the station involves levelling from the nearest Ordnance Survey bench-mark, or from the nearest point for which the altitude is accurately known. In fairly flat country, observers should be able to determine the altitude of the station without much difficulty by reference to a large-scale map on which altitudes are shown. In more difficult situations it is better to consult a surveyor.

The reference datum from which heights above mean sea level are measured is Newlyn for stations in mainland Great Britain; for stations in Northern Ireland it is Belfast, but occasionally the Irish Dublin Bay datum reduced to Newlyn is used; for Shetland the Lerwick datum is used, and for the Western Isles and other islands some distance from the mainland the appropriate local datum is used.

Values found in feet should be converted to metres by multiplying by 0·3048, then rounding off to the nearest metre.

I.4. ORIENTATION

For the determination of wind direction, and for some other purposes, it is necessary to know the exact direction of true north from the station and the true bearing of selected conspicuous objects. Several alternative methods are available for doing this.

(a) A magnetic compass may be used for determining the orientation of a station, provided careful precautions are taken. A compass needle does not point to the true geographic pole but diverges to west or east of north by an amount, known as the 'declination', which differs from place to place and, in any one place, varies slowly from year to year. All directions determined by compass bearing must be corrected by the amount of the magnetic declination, which may be obtained by consulting the surveying or engineering authority for the town and area. The bearings are obtained from magnetic bearings by subtracting a westerly declination or adding an easterly one.

In the British Isles at the present time the magnetic needle points west of true north. An isogonal chart showing approximately the amount of declination or variation (for epoch 1977·5) is given in Figure 21. This chart is not reliable for certain places where the magnetic field is disturbed by local effects. Recent local information should be used, if available, in preference to the values shown.

A serious source of error in the determination of direction by means of a magnetic compass is the disturbing effects produced by the proximity of iron or steel or of electric currents. Vehicles and iron railings are major disturbances but even the presence of such small objects as iron nails in the support on which the compass is placed, or a knife or keys in the observer's pocket, may cause serious errors of unknown magnitude. The observer must satisfy himself that all such possible sources of disturbance are absent before a compass reading can be accepted as reliable.

(b) The Ordnance Survey map may be used to obtain true bearings from the station by using conspicuous objects in the neighbourhood such as church steeples or prominent points in the landscape. For accuracy it is advisable to select the most distant object possible within the limit of the visual range.

On the map, when a suitable distant object has been identified, carefully draw a straight pencil line from the station to the object and, if necessary, extend the pencil line to cross a north–south grid line. Measure any of the two angles between the pencil line and the north–south grid line with a protractor and convert the angle obtained to a bearing relative to 'grid north'. As 'grid north' usually differs slightly from true north by an amount which depends on the district, true north is obtained by adjusting the grid bearing by the amount printed on the map.

(c) Another method may be used by stations in the northern hemisphere, based on the position of the pole-star. This star marks the north point within about one degree and may be easily identified on any clear night; the plane of the meridian, i.e. the north–south plane, passes approximately through the pole-star and through the zenith and the observer. For accurate results, or for using the same method with other stars, as is necessary in the southern hemisphere, the information given in the *Nautical almanac* may be employed. Observe a known star through a theodolite and set the azimuth scale so that it reads the azimuth obtained from the *Nautical almanac* for the chosen star at the time of observation. The zero of the scale will then read true north. If the theodolite is left in position overnight, the bearings of conspicuous objects can be obtained during daylight.

(d) The direction of true north can be determined to a high degree of accuracy by using the shadow cast by a vertical object when the sun is due south at the time of local noon. This method does not require a knowledge of the magnetic declination or of local topography; only the longitude and correct time need be known.

The exact time GMT that corresponds to local noon at a particular longitude, east or west of Greenwich, can be deduced from Table III in Appendix III.

Having synchronized a clock or watch against a reliable time-signal and calculated the time of local noon from Table III, the north–south plane can be obtained by marking the shadow cast at that time by any vertical object such as a flagpole, a corner of a building, etc.

A theodolite is then set up over this line and adjusted to read zero degrees when pointing north. The theodolite can then be used to find the true bearing of any object required.

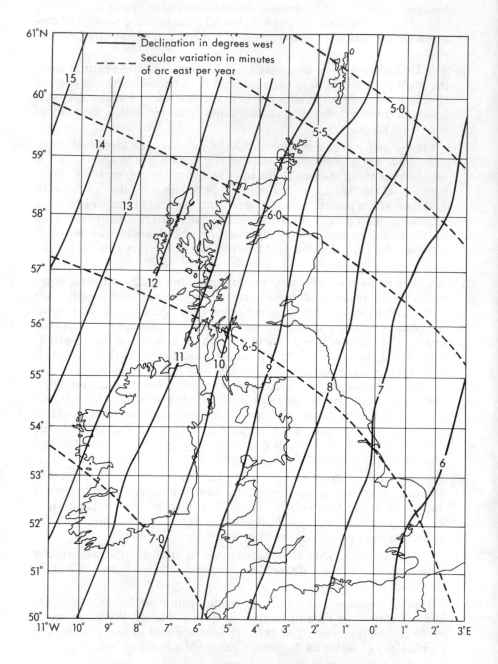

Figure 21.   Isogonals or lines of equal westerly declination,
1977·5 (1 July 1977)

The map was supplied by the Geomagnetism Unit, Institute of Geological Sciences, for the epoch
1977·5 and includes contour lines of equal values of the annual rate of change of the declination.

I.5. INSTRUMENTAL EQUIPMENT

The essential instrumental equipment for a synoptic and a climatological station comprises:

    Thermometers, dry-bulb and wet-bulb
    Thermometers, maximum and minimum
    Thermometer screen
    Rain-gauge and rain measure.

Other instruments essential for a synoptic station and sometimes used at a climatological station are:

    Barometer
    Barograph
    Anemometer and wind vane (or anemograph)
    Grass minimum thermometer (essential at agrometeorological stations).

Desirable additions to this equipment include the following:

    Sunshine recorder (essential at agrometeorological and Health Resort
        stations)
    Soil thermometers (essential at agrometeorological stations)
    Concrete-slab minimum thermometer
    Thermograph
    Hygrograph
    Recording rain-gauge
    Bare-soil minimum thermometer
    Cloud searchlight and alidade ⎫
    Pilot-balloon equipment   ⎬ essential at stations on aerodromes.
    Visibility equipment   ⎭

A number of other items not listed may be necessary or desirable at certain stations, for example a cloud-base recorder at aerodromes, a cup-counter anemometer at agrometeorological stations. The instrumental equipment at an auxiliary reporting station is based on the needs of the Meteorological Office, the suitability of the site and the availability of observers.

Precision aneroid barometers are certified by the Meteorological Office and mercury barometers by the National Physical Laboratory. All types of thermometers are verified by the Meteorological Office and those used for routine meteorological observational purposes are no longer issued with certificates unless required for a special purpose for which the application of corrections is justified. The accuracies of rain-gauges, rain measures and sunshine recorders are also verified before being issued to stations. Arrangements can be made by the Meteorological Office for such instruments, if purchased privately, to be examined and verified for accuracy on payment of a fee. Details of this scheme can be found in Met.O. Leaflet No. 9: Meteorological Services — Testing and inspection of meteorological instruments.

Brief details of recommended types of instruments, with notes on exposure and installation, are given below.

I.6. BAROMETERS

There are two types of barometer approved by the Meteorological Office: the precision aneroid and the Kew-pattern mercury barometer, as described in

Chapter 7. The precision aneroid has largely superseded the Kew-pattern barometer at Meteorological Office and auxiliary reporting stations.

I.6.1. Precision aneroid.   The barometer should not be installed in a position where it could be subjected to knocks, jolts or rapid variations of temperature. It must not be exposed to direct sunshine. Where a barometer has to be installed in an air-conditioned office or in a room in a high building, the advice of the Meteorological Office should be sought as to the advisability of using a static pressure head to improve the exposure (see 7.2.5, page 102).

I.6.1.1. *Installation.*   The barometer must be installed with the base of the instrument horizontal. A Mk 2 barometer can be mounted on a wall by means of a mounting plate or, if a bench mounting is preferred, housed in the wooden carrying case which should be attached firmly to a horizontal bench. A Mk 1 barometer must only be bench- or shelf-mounted; the calibration of this type of instrument will be disturbed if it is fixed to a mounting plate. Under no circumstances must the instruments be left free-standing. The height of the instrument above floor level should be approximately 1·2 m.

When an instrument is to be mounted on a wall, it is first necessary to fix two horizontal wooden battens to the wall. These should be approximately 170 mm long, 25 mm wide and 15 mm thick and be placed about 110 mm apart between the inside edges. If two barometers are to be installed side by side the battens should be 550 mm long. The instruments should be fixed towards the ends of the battens. The battens must be clear of any return wall. If the wall is not solid the battens must be long enough to be fixed to firm uprights. From the back of the mounting plate insert three 2 BA screws through towards the front, using the washers and thin 2 BA nuts supplied; then screw the mounting plate to the battens, using the wood-screws and spacers supplied, the spacers going between the battens and the plate. Remove the barometer from its carrying case by unscrewing the knurled captive retaining bolts which pass through the bottom of the case. Take the triangular back-plate from the bottom of the carrying case and, using the screws supplied, screw it to the back of the barometer with the three spacers between the barometer and the plate. Position the barometer so that the three screws protruding from the mounting plate pass through the back-plate. To keep the barometer in this position, use the 2 BA stiff nuts and tighten them firmly.

If bench mounting is necessary, the barometer and its carrying case must be treated as one unit. The barometer must be secured in its case by the knurled retaining bolts through the base before the case is fixed to the bench because the bolts would be inaccessible once the case was fixed. The bottom of the case is drilled to take three No. 8 wood-screws. A pair of pliers or tweezers will be needed to insert two of the screws through the base on the left-hand side of the case. The three wood-screws should then be screwed down firmly to the bench.

I.7. BAROGRAPH

A small barograph is suitable for use at climatological stations where the barograph is a non-essential instrument, but the large pattern must be used at synoptic stations in order that pressure tendencies may be determined with the desired accuracy.

Some notes on exposure are given in 7.7.1 (page 107). A shelf near the barometer usually provides a suitable position. If the building is subject to vibration, the barograph should stand on a pad of sponge-rubber or felt.

I.8. THERMOMETERS AND SCREENS

Ordinary thermometers used for measuring dry-bulb and wet-bulb temperatures in a naturally ventilated screen, and maximum and minimum thermometers, must be manufactured in accordance with British Standard Specification 692: 1976. The exercising of control by the Meteorological Office in the testing of all thermometers dispenses with the need for certificates for scale corrections, the thermometers having been tested for compliance with the appropriate British Standard or Meteorological Office specification before issue. The tolerances allowed in these specifications vary from type to type but are within the accuracies normally required for routine observing purposes.

The five different types of thermometer required to provide all the thermometric observations described in Chapter 8 are:

Ordinary thermometers for use as dry-bulb and wet-bulb
Maximum thermometer
Minimum thermometers for determining screen minimum, grass minimum, bare-soil minimum and concrete-slab minimum temperatures
Soil thermometers, one type for depths of 30 cm or more, and one type for depths of less than 30 cm.

I.8.1. Sheathed thermometers. The sheathed pattern standardized for maximum, minimum and ordinary thermometers has several advantages. Two of the more important are:

(*a*) The glass sheath protects the graduations and figuring on the stem from the weather, so that they remain permanently clean and legible.

(*b*) Sheathed thermometers are much more robust than solid-stem thermometers.

I.8.2. Range of temperature scales. The ranges to be covered by the scales of thermometers for use in temperate latitudes and the tropics are as follows:

	Temperate latitudes	Tropics
	degrees Celsius	
Ordinary	−30 to 45	−20 to 55
Maximum	−20 to 55	−10 to 65
Minimum	−35 to 40	−25 to 50

I.8.3. Thermometer screens. The standard exposure for the measurement of air temperature is provided by a double-louvered box or screen, originally designed by Thomas Stevenson. This is erected in an open situation so that the bulbs of the wet- and dry-bulb thermometers are 1·25 m above the ground. As far as possible the ground cover beneath the thermometer screen should be short grass, or the natural earth surface of the district at places where grass does not grow.

Two versions made of wood are in general use:

(a) Ordinary thermometer screen designed to hold dry- and wet-bulb, maximum and minimum thermometers.

(b) Large thermometer screen (Plate XXV, opposite page 124), similar to (a) but made long enough to accommodate a thermograph and hygrograph as well as the usual four thermometers.

Galvanized metal stands are available for both of these types of screen; they are much more durable and satisfactory than wooden stands, though the latter may be used.

No additional shelter such as a roof erected above the screen is allowed, even in the tropics.

I.8.3.1. *Installation of thermometer screen.* The stand should be assembled and stood upon the site (chosen in accordance with I.2) with one of the longer sides facing towards north. With the base of the stand as a guide, the shape of the hole to be dug should be marked in the turf; the hole should be longer than the base of the stand by about 30 cm each way. Move the stand away and dig out a hole 30 cm deep, saving the turf for subsequent replacement. Place the stand in the hole and carefully check it for level and orientation. Partially fill in the hole, place the screen temporarily in position on the stand and check again. Shovel in the rest of the soil, ramming it down well, and replace the turf. Place the screen on the stand with its door opening towards the north, then screw it into position. (This orientation is to prevent, as far as possible, the sun shining upon the bulbs of the thermometers at any time when the door is opened.) If the ground is soft so that there is a risk of irregular settlement, it is best to dig the hole a few centimetres deeper and put in hard core, well rammed, to give the feet of the stand a firm foundation. The legs may have to be set in individual blobs of concrete at windy sites. When finally installed, the base of the screen should be 1·1 m above the ground.

I.8.4. Soil thermometers for depths of 30 cm or more. These are installed in steel tubes of the requisite depth driven vertically into the ground. From Figure 9B (page 111) it will be seen that in tubes of the Meteorological Office pattern the base of the outer glass case of the thermometers rests on a rubber pad just above the conical point of the tube. Near the top of the steel tube there is a flange which is so placed that when the flange is in contact with the surface of the ground the thermometer bulb is at the correct depth. The upper end of the steel tube is closed by means of a polythene cap to prevent the entrance of rain-water and dirt.

Before installing the steel tube, first drive a pilot hole so that the tube can then be inserted without excessive hammering. A block of wood should be placed over the end of the tube to take the blows of the hammer or mallet, and care should of course be taken to ensure that the tube is kept vertical.

The normal depths for such thermometers are 30 cm and 100 cm but, exceptionally, the temperatures at other depths may be measured. The thermometers should be installed under a grass-covered surface. The surface should be uniform and horizontal in all directions to a distance at least equal to the depth of the deepest thermometer bulb. The site should be well exposed to sunshine and the grass should be kept weed-free and reasonably short because long grass retains an undue amount of moisture. In the event of appreciable snowfall, the snow should be carefully removed from the tops of the caps before the thermometers are read and afterwards replaced as far as possible.

Where two tubes are installed they must be separated by a distance equal to the length of the longer tube.

I.8.5. Soil thermometers for depths of less than 30 cm.   These thermometers are very fragile and should be handled with great care (see Figure 9D, page 112). At meteorological stations, where normal depths of measurement are 5, 10 and 20 cm, the thermometers are installed in bare soil. To make the holes, use a rod of the same diameter as the bulbs of the thermometers, choosing an occasion when the soil is moist and firm. Make the holes slightly deeper than necessary and put a little sifted soil into each so that, when the thermometer is inserted, the bulb rests on this soil when the horizontal graduated part of the stem is in contact with the ground. Fill in the holes with sifted soil, packing it down gently with a thin piece of wood. The aim should be to disturb the soil as little as possible. When each thermometer is put into its hole make quite sure that the mercury thread is free from breaks.

Precautions regarding exposure, keeping the soil free from weeds, readings after snowfall, and homogeneity of the surface are similar to those described in I.8.4 above. There should be little or no protection of the thermometers, although a wire not more than about 30 cm high stretched round firm pegs at the corner of the bare patch is permitted to prevent accidental damage.

After a period of time, because of compaction of the soil by rainfall, it may be found that the horizontal stems of these soil thermometers have become clear of the surface and the depth of the bulb is correspondingly reduced. This source of error should be corrected by adding sifted soil to the surface around the thermometers until their horizontal stems are again just in contact with the ground.

I.8.6. The grass minimum thermometer (Figure 9E, page 112) is exposed over short grass, supported either on two forked pieces of wood obtained locally or on the special grass minimum thermometer supports. The wooden pegs should support the thermometer so that the bulb is between 2·5 and 5 cm above the ground and in contact with the tips of the grass blades. The bulb of the thermometer should be a little lower than the other end, and should be well clear of the nearer wooden support. If the special rubber supports are being used, by choosing a particular side of the squares to rest on the ground the thermometer bulb will be just touching the tips of the grass and the instrument will be at the necessary inclination to the horizontal.

To minimize the risk of accidental damage to the grass minimum thermometer, it is desirable to make its position more conspicuous by using white-painted pegs as corner posts to a miniature fence around it. No form of wire screen or other protective device should be placed over the thermometer.

I.8.7. The bare-soil minimum thermometer is laid on level bare soil with the stem having a very slight slope downwards towards the bulb; the bulb must

G

never be higher than the stem. Small pegs of wood should be placed in the ground at each side of the thermometer, but not near to the bulb, to prevent accidental movement.

The bare plot in the enclosure (see Figure 20) is a suitable site. The same site, once selected, should always be used.

I.8.8. The concrete-slab minimum thermometer is exposed in the centre of a concrete slab or flag in natural colour and measuring 1 m by 0·6 m and 5 cm thick, conforming to British Standard 368: 1971. Such slabs are usually readily available but, if one has to be made up locally, it must conform to the above standard: no alternative is acceptable.

The slab should be laid horizontally with its smooth side uppermost and flush with the ground in as open a position as possible. In practice, the position selected should be in the enclosure, at least 1·5 m from the rain-gauge and at least 1·25 m from the nearest fence and all other instruments. A suitable position is indicated in Figure 20.

At the chosen position the turf should be skimmed off from an area equal to the size of the slab and to a depth of a little more than 5 cm. The bare earth should then be covered with a layer of sand to such a depth that the concrete slab, when placed in position and bedded down, should have its smooth horizontal surface very slightly higher than the surrounding grass surface. This is to allow for settlement and to prevent the collection of more than a very thin layer of water on the slab during rain.

The slab must be drilled and plugged to take a 1-inch No. 6 brass wood-screw so that a PVC-coated spring clip can be fixed to hold a grass minimum thermometer in in position. The thermometer is placed on the slab with its bulb in contact with the centre of the concrete and its stem parallel to the longer side of the slab. The spring clip is positioned so that when the thermometer is in place the end of its anti-condensation shield just touches the clip. The thermometer will then have the correct downward slope towards the bulb of about 2°.

I.9. RAIN-GAUGE AND RAIN MEASURE

I.9.1. The rain-gauge should be somewhere near the centre of the enclosure and not less than 3 m away from the screen. It should also be at a distance of not less than twice the height of any pillar used in the enclosure to support a sunshine recorder or fencing post. At stations where there is a requirement for a second rain-gauge, it should be placed in the most convenient position which satisfies the rules for rain-gauge exposure. (Figure 20 shows one possible position.) If for any reason the rain-gauge cannot be installed in the enclosure, it should be sited so that its horizontal distance from any surrounding object is not less than twice the height of the object above the rain-gauge. Provided these rules are adhered to, the presence of objects tending to shelter the rain-gauge from the wind is advantageous rather than the reverse. For rain-gauges which are installed in forest openings, tree heights should be less than half the distance from the forest edge to the rain-gauge.

The Meteorological Office standard rain-gauge Mk 2 (made mainly of copper and with a collecting funnel 5 inches (127 mm) in diameter) is sunk firmly into the ground so that the rim of the funnel is horizontal and 30 cm above the ground.

A circle of small stones or granite chippings for a few centimetres round the base of the gauge reduces the possibility of damage to the gauge during grass cutting.

I.9.2. Turf wall. In very open situations the catch of the gauge may be seriously reduced by eddies set up by the gauge itself when the wind is strong. In such situations it is necessary to set up a turf wall round the gauge (Figure 22). This consists of a sloping circular embankment, the crest of which is horizontal and in the same plane as the rim of the gauge. To construct a turf wall for a 5-inch copper rain-gauge the first step is to erect a low circular brick wall or wooden fence enclosing an area 3 m in diameter with the gauge in the centre. Drainage must be provided to prevent water accumulating and forming a puddle inside the fence. Earth is built up outside the fence to form a surface sloping downwards with a gradient of about 1 in 4; it should be made firm and finished off with turf. Care should be taken to maintain a turf wall in good condition. Figure 22 shows the dimensions of a wall for a standard copper gauge with the rim set 30 cm above the ground.

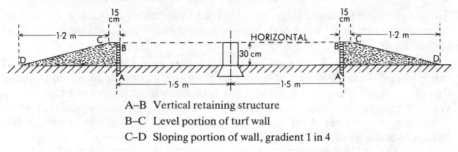

A–B   Vertical retaining structure
B–C   Level portion of turf wall
C–D   Sloping portion of wall, gradient 1 in 4

Figure 22.   Turf wall for use at exposed rain-gauge sites

I.9.3. Recording rain-gauges (see 9.5, page 144) require the same sort of exposure as ordinary gauges. A turf wall, as described in I.9.2, may be constructed around the Meteorological Office tipping-bucket gauge. However, it is usually not practicable to provide such shelter for the Meteorological Office tilting-siphon rain recorder because its rim is set 52 cm above the ground.

If the water-table rises within about 10 cm of the surface of the ground, the bottom of the siphon tube of a tilting-siphon recorder may dip into the water, thus preventing the recorder functioning properly. This may happen on occasions of very heavy rainfall and can be completely prevented only by expensive piped drainage. The trouble may, however, be avoided (except on the most troublesome sites on occasions of exceptional rainfall) by installing the gauge on the edges of two concrete slabs set in a deep bed of porous rubble.

It should be noted that conventional recording gauges are mainly required to provide information about the time, duration and rate of precipitation. They do not replace the ordinary gauge as a reliable instrument for measuring the total fall.

See 9.5.1.4, page 146, concerning precautions in frosty weather when using a tilting-siphon rain recorder.

I.9.4. Rain measures should be of the tapered pattern (see Figure 13, page 139), certified by the Meteorological Office and graduated in millimetres. A flat-based measure may be used for large-capacity gauges.

I.10. SUNSHINE RECORDER

I.10.1. Exposure. The sunshine recorder should be set up on a firm and rigid support, ideally in a place where there is no obstacle to obstruct the sun's rays at any time of the day at any time of the year.

In the British Isles a free horizon is required in the approximate ranges north-east through east to south-east, and north-west through west to south-west. Any obstruction to the south should not be allowed to cut off sunshine during the period of lowest midday elevation of the sun in December, and this condition will be met if the distance of the object and its height above the sunshine recorder are related as follows:

Latitude 50°N: distance at least 3½ times the height.

Latitude 55°N: distance at least 5 times the height.

Latitude 60°N: distance at least 9 times the height.

These ratios are in round figures and refer to objects due south of the instrument. Obstruction to the north between north-west and north-east is of no consequence.

More precise information about the exposure requirements for these and other latitudes may be derived from the diagrams in Figure 23. Each of these diagrams shows the altitude and azimuth of the sun at different times of the year for a particular latitude; the hours of the day are given in Local Apparent Time. The five curves A to E on each diagram are for different times of the year according to the following key:

Date	Sun's declination	Hemisphere Northern	Southern
22 June	23½°N	A	E
21 April, 23 August	11¾°N	B	D
21 March, 23 September	0°	C	C
18 February, 25 October	11¾°S	D	B
22 December	23½°S	E	A

A = summer solstice   C = equinoxes   E = winter solstice

The final diagram of Figure 23 is for a latitude of 30° in the southern hemisphere. This has been included to demonstrate the symmetrical interchange which must be effected in the hours and azimuths if the various latitude diagrams for the northern hemisphere are to be used at southern hemisphere locations.

The sun does not attain sufficient power to scorch the card until it is about 3° above the horizon. Therefore obstacles which subtend an angle of not more than 3° do not cause loss of record. As a rough guide, an angle of 3° is that subtended by an object 10 m high and 200 m away, or an object 100 m high and 2 km away. In applying this criterion, the height of the object is taken to be the amount by which the height of the object exceeds that of the recorder.

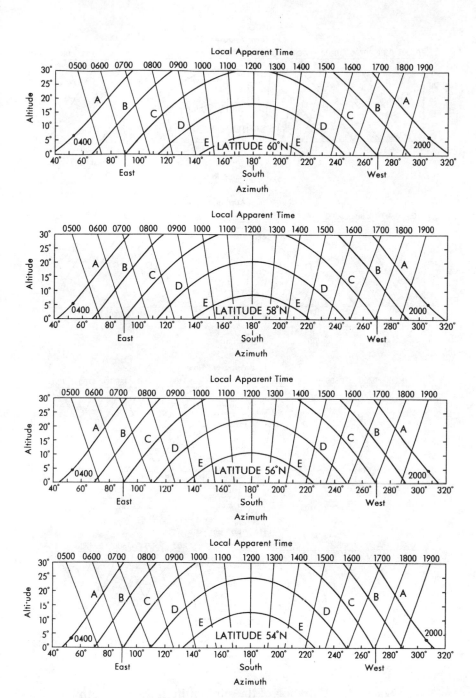

Figure 23.   Variation of the sun's altitude and azimuth
For explanation see page 190
Latitudes 60°–54°N

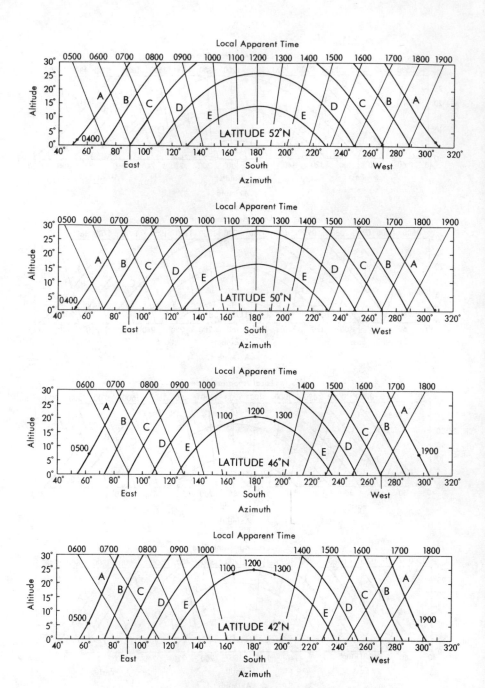

Figure 23.   Variation of the sun's altitude and azimuth
(*continued*)
For explanation see page 190
Latitudes 52°–42°N

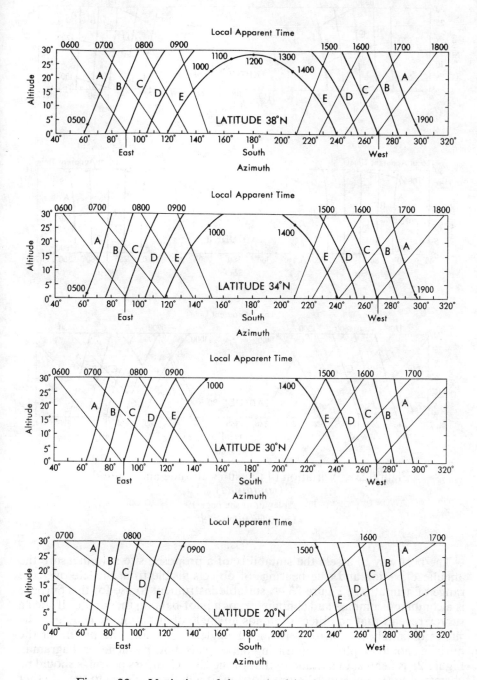

Figure 23.   Variation of the sun's altitude and azimuth
(*continued*)
For explanation see page 190
Latitudes 38°–20°N

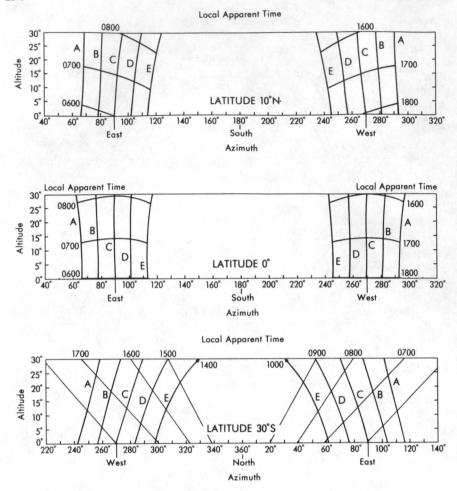

Figure 23.   Variation of the sun's altitude and azimuth
(*continued*)
For explanation see page 190
Latitudes 10°N–30°S

The best way to assess the suitability of a proposed site is to measure the angular elevation and true bearings of objects visible from the site over the range of azimuths in Figure 23. A suitable instrument to use for this purpose is a combined compass and clinometer, or a pilot-balloon theodolite. If such a survey reveals that no object subtends an angle of elevation exceeding 3° the site is definitely satisfactory. If higher angles are found, the profile of the horizon should be plotted on tracing paper placed over whichever diagram in Figure 23 is nearest in latitude to that of the site. Compass bearings should be corrected to true bearings before plotting. It will then be easy to see whether the obstacles will or will not obstruct the sun's rays at any time of year.

If the obstructions are near at hand, the site may be rendered satisfactory by installing the recorder on a brick pillar or other rigid support high enough to provide clearance. To avoid obstruction, however, it is often necessary to install the recorder on the roof or parapet of a building, but positions near chimneys from which smoke may issue should be avoided.

It may be found that even at the best available site there are obstacles which cause loss of recordable sunshine. Within certain limits, records from such sites may be accepted by the Meteorological Office.

I.10.2. Installation. When a satisfactory exposure is available near ground level the recorder may be installed on a brick or concrete pillar. If this pillar is less than 2 m high it may be sited in the instrument enclosure (see Figure 20). If necessary, steps should be built beside the pillar to enable the observer to change the card. Any such installation of sunshine pillar and steps must not over-shelter the rain-gauge(s). Hard bricks impervious to rain should be used.

The platform on which the instrument is fixed must be perfectly rigid and not liable to warp or be otherwise damaged by weather. A suitable platform can be made from a block of cement 30 cm × 30 cm and not less than 5 cm deep in which three fibre plugs are embedded to take the screws which hold the base of the recorder in position. It is best to determine the positions for the plugs first by moving the instrument bodily in azimuth until the meridian adjustment (see I.10.3.2) is approximately correct. This should be done with the levelling screws at the midway position of the slots in the sub-base, and with the sub-base level. The plug holes may then be drilled and the base screwed down ready for the final adjustments to be made.

I.10.3. Adjustment. Before delivery, the recorder is adjusted for concentricity, that is to say the centre of the sphere is made to coincide with the centre of the frame. This adjustment should on no account be disturbed on a Mk 3 recorder, but a centralizing gauge can be used on a Mk 2 recorder.

Adjustments have to be made after the recorder has been installed to ensure that the image of the sun will cross the noon line exactly at 1200 hours Local Apparent Time (refer to Table III on page 208 to find the time GMT) and that at the equinoxes (21 March and 23 September) the trace will coincide with the horizontal central line of crosses of the equinoctial card. If these conditions are fulfilled the recording will be correct at all times of the year, and then it can be said that:

(*a*) the instrument is level in the east–west direction,

(*b*) the instrument faces due south (north in the southern hermisphere), that is when the plane passing through the centre of the sphere and the noon marks on the bowl is in the plane of the meridian, and

(*c*) the plane passing through the centre of the sphere and the central line of the equinoctial card is inclined to the vertical at an angle equal to the latitude of the place.

In addition to these initial adjustments, warping or subsidence of the support may make further adjustments necessary; these can be made provided that the surface of the support on which the recorder is mounted is still reasonably level.

I.10.3.1. *Adjustment for level.* Slacken off the three lock-nuts of the levelling screws. Place an accurate spirit-level across the horns of the bowl, taking care to see that the axis of the level is exactly east–west. This is best done by inserting a card in the recorder and making the spirit-level touch the card at either end. Bring the bubble to the centre of the level by adjusting the two front levelling screws. Then place the level on the sub-base in the north–south direction and adjust by means of the rear levelling screw only.

I.10.3.2. *Adjustment for meridian.*   This adjustment can be made satisfactorily only when the sun is shining and is best carried out in the middle of the day at an exact hour of Local Apparent Time. Table III on page 208 shows the time, GMT, at which the sun's image should appear on the noon line of the card on given dates for each degree of longitude between 4°E and 10°W. The times for intermediate dates and longitudes can readily be found by interpolation. For longitudes outside the range covered by the table the required times can be found by taking the values tabulated under longitude 0° and adding (subtracting) 4 minutes for every degree west (east) of Greenwich.

The procedure is as follows. First set your watch accurately to Greenwich Mean Time by means of, say, a radio time-signal (taking into account the difference between summer time, when in force, and Greenwich time). From Table III obtain the time of local apparent noon for the required longitude and date. Insert a card of the right type for the season into the recorder and set it accurately to the noon marks. With the locking nuts of the levelling screws slackened off, move the sub-base as necessary so that the sun's image is exactly on the noon line of the card at the time, GMT, evaluated from the table.

The differences shown in Table III between Local Apparent Time and GMT are equally applicable to any other hour of the day. For example, at a station in longitude 4°E on 1 July the sun's image should be on the 1100 line at 1047 GMT, on the 1300 line at 1247 GMT, etc.

I.10.3.3 *Adjustment for latitude.*   The bracket which supports the bowl of the standard recorder is slotted and provided with a scale of degrees, so that if the latitude index is set to read the latitude of the place and the sub-base is level in the north–south direction, then requirement (*c*) of I.10.3. is met. To test the adjustment, the position of the trace should be compared with Figure 17 (page 159) which shows where the trace should come on any given date. For example, on 11 March or 3 October the trace should be 5 mm above the centre line of the equinoctial card, in the northern hemisphere. On 28 April or 15 August the trace should be exactly on the central line of the summer card. If such a comparison shows an appreciable error it is best corrected by an adjustment of level in the north–south direction, using the rear levelling screw.

When all adjustments have been made the lock-nuts should be tightened and the adjustments finally checked.

I.11. WIND VANES

Where a remote-transmitting wind vane is not installed it is highly desirable to provide a wind vane conforming to the requirements in 5.3.1 (page 84), preferably of the standard Meteorological Office pattern described therein.

I.11.1 Exposure. Whichever type of vane is selected it is essential that it should be satisfactorily exposed. In an open situation with no obstacles such as trees or buildings in the near vicinity the vane may be erected on a metal mast clamped to a stout wooden post so that the vane is about 6 m above the ground. Where there are obstructions, however, the eddies they create will cause excessive oscillations of the vane, making the wind direction difficult to determine. This difficulty may be overcome by increasing the height of the mast so that it is at least 3 m higher than any obstacle in the vicinity. If this would necessitate a very high mast in the vicinity of the enclosure, a separate site should be selected for the wind vane.

I.11.2. Installation. Matters of major importance in erecting the vane are, firstly, the spindle about which the vane turns must be truly vertical, and secondly, the arms indicating the cardinal points must be correctly orientated to true (not magnetic) directions.

Methods of determining the directions of true north are described in I.4 (page 180). The direction arms are attached to a boss which can be clamped in position by means of a nut. This adjustment may be made by sighting along the north–south line after the mast has been erected and made vertical. The base of the Mk 2B wind vane has a 1½ inch BSP socket for screwing to a steel pipe or mast.

I.12. ANEMOMETERS

Brief descriptions of anemometers used by the Meteorological Office will be found in 5.4 (page 88). Though not an essential instrument at climatological stations in general, an anemometer is a desirable addition to the equipment if the station has been set up for special purposes such as agricultural meteorological studies or industrial research. Cup anemometers of either the generator or contact pattern (for observations at fixed times) or the counter pattern (for measuring the run of wind over extended periods) will meet most requirements at such climatological stations.

Where continuous records are necessary an anemograph should be installed. As these are costly instruments, authorities are advised to consult the Meteorological Office regarding siting etc. as soon as installation is contemplated.

I.12.1. Exposure. For synoptic and climatological purposes the standard exposure for an anemometer over open terrain is at 10 m above the ground. Open terrain may be defined for this purpose as level ground with no obstruction within 300 m. The standard exposure is not often obtainable in practice and, where these conditions cannot be met, adjustments should be made to the height of the anemometer so that wind speeds and directions can be obtained which are generally representative, as far as possible, of what the wind at 10 m would be if there was no obstruction in the vicinity. It is difficult to give any brief general rules for determining the height required because local conditions differ so widely. The notes below are a general guide on exposure requirements, but advice regarding particular situations may be obtained from the Meteorological Office Climatological Services Branch.

At a site where any obstructions are not large and are distributed more or less uniformly around and in the immediate vicinity of the anemometer, to give an effective height of 10 m the anemometer needs to be at a height $H +$ 10 m, where H is the approximate height of the tallest of the various obstacles.

At those locations where there are large obstacles (e.g. 12 m or more in height) within 150–300 m, it will be necessary to raise the anemometer to such a height that the wind reaching the anemometer, after passing over such obstacles, is affected by them as little as possible and excessive gustiness avoided. As a rough guide, the minimum height of an anemometer above ground when there are obstacles of height h (where h is 12 m or more) at various distances around it is given by the following table:

Distance of obstruction	Minimum height of anemometer
h	$2{\cdot}0\,h$
$5\,h$	$1{\cdot}67\,h$
$10\,h$	$1{\cdot}5\,h$
$20\,h$	$1{\cdot}25\,h$
$25\,h$	$1{\cdot}13\,h$
$30\,h$	h

Thus, in the case of a building 15 m high at a distance of 300 m (20 times the height of the building) from the proposed site of the instrument, the anemometer would need to be about 19 m (i.e. 15 m $\times$ 1·25) above the ground.

The above guidance must be used with some discretion. To increase the height of exposure of an anemometer because of a single isolated obstruction with small horizontal dimensions compared with its height will then mean that the data from the much larger unobstructed sector will require correction to the standard height of 10 m. In a similar fashion, where there are more numerous obstacles which may not be regular either in height or distribution around the proposed anemometer site, the effective height may differ appreciably for winds from different directions and this makes correction of the speeds to a standard level difficult. Such sites should therefore be avoided.

When the anemometer needs to be mounted on an isolated building, the building itself will disturb the wind flow to some extent, depending among other things on its size and shape. As a rough guide, a mast or tower erected on the roof needs to be at least half, and probably about three-quarters, of the height of the building, assuming that the building has considerable horizontal extent (which excludes such things as water towers, lighthouses, etc.) Thus, with a building 12 m high, a roof-top mast at least 6 m and preferably 9 m high should be used. The effective height in such a case may be taken as about the height of the mast plus half the height of the building, i.e. 12 or 15 m in the case quoted.

For the records to be of value, changes in the site exposure should be kept to a minimum. The subsequent erection of buildings in the vicinity of the anemometer site can affect the data recorded, sometimes so seriously that the site becomes useless for wind measurements. Slow changes in exposure, due to the growth of trees, are often not readily identifiable from the data because of the natural variability of the wind. When installing an instrument it is therefore advisable to make a detailed site plan showing the height and extent of all obstacles within 300 m. The plan should be reviewed every few years, the changes of exposure noted and, if possible, action taken to minimize any obvious trends.

Full details of installation procedures are available from the Operational Instrumentation Branch.

APPENDIX II

RECORDING OF OBSERVATIONS FOR CLIMATOLOGICAL PURPOSES AT OBSERVING STATIONS

The Meteorological Office is the principal national custodian of meteorological data, and as such is called upon by many interests to supply information or advice based upon climatological considerations, e.g. for aviation, agriculture, hydrology, engineering, industry. Moreover, reliable meteorological data are required for the study of changes of climate. The climatological returns from all stations are thus of great value and of considerable economic importance. Returns are often photocopied for answering enquiries and the data entered on them are processed directly to a computer-accessible form; thus writing must be legible, and in black or blue-black permanent ink. Blue and other light shades of ink do not photograph well. Inks which fade with time should not be used. On no account should pencil be used.

II.1. CLIMATOLOGICAL STATIONS

The observations made at a climatological station reporting to the Meteorological Office are recorded in two ways. The original observation is entered in a Register (Metform 3100) at the time it is made: this Register is usually retained at the station. These observations are then transcribed to a form which constitutes a return. A return is completed for each month (Metform 3208B) and, in special cases, each week (Metform 3110/A/B). These monthly and weekly returns are sent to the Meteorological Office where they are used in the preparation of official weather reports and climatological summaries. They are filed at the Meteorological Office as the permanent register of the daily weather at the station concerned.

Climatological stations co-operating in the Health Resort Scheme also make returns of climatological observations on Metform 3208B. These returns are in addition to the special daily telephoned reports, details of which are supplied separately to those stations involved.

II.1.1. Metform 3100: Pocket Register for observations at climatological stations. The headings of the columns in this Register are largely self-explanatory.

Temperatures, wind speed and direction, rainfall and sunshine are among the entries which are made in a straightforward manner. Visibility is recorded in the letter (or number) scale which is directly derived from a series of visibility objects, and the state of the ground is recorded on one of the two scales given in 6.2 (page 96). The last column of the Register is headed 'Weather diary and remarks' and leaves sufficient space for valuable additional entries on unusual phenomena, the timing of particular events, a summary in Beaufort letters of the weather of the day ending at midnight GMT, and so on.

It should be clearly stated in the Register if Greenwich Mean Time or some other standard of time is being used; the standard having been stated, it should be strictly adhered to for all purposes.

Two special points are sometimes overlooked: firstly, when there has been no precipitation the appropriate entry in the Register should be a dash (' — ') and not '0·0'; secondly, when there has been no sunshine the entry should be '0·0'.

In both these cases, as with other types of observation (temperature or wind, for example), if no observation has been made the appropriate space should be left blank and a qualifying remark, giving the reason whenever possible, should be entered in the 'remarks' column.

Correctness in all these matters when using the Pocket Register will greatly simplify the preparation of monthly and weekly returns. Climatological observations entered in the Pocket Register are not normally required immediately, unlike synoptic observations which must be coded and transmitted at the regular hours of observation. Regularity and promptness are no less important in maintaining the record, if the full value of the observations is to be assured.

II.1.2. Monthly returns at climatological stations: Metform 3208B. Detailed instructions for the completion of Metform 3208B at co-operating climatological and agrometeorological stations are given in Metform 3100A which is the supplement to the Pocket Register and is issued to all observers concerned.

II.1.3. Weekly returns at climatological stations: Metforms 3110/A/B. Certain stations make weekly returns; Metform 3110 is used in England and Wales, Metform 3110A in Scotland and Metform 3110B in Northern Ireland, and they are dispatched to Meteorological Offices in Bracknell, Edinburgh and Belfast, respectively. The completed forms are used for the preparation of official reports to the Registrar-General and for other purposes. The headings and notes on the form call for little explanation and the information required can, for the most part, be taken directly from the Pocket Register when this is correctly maintained.

Metform 3110 is divided into two sections, one for completion at co-operating climatological stations, the other for use at meteorological offices and at auxiliary reporting stations (see II.2.3).

All totals and means should be calculated and checked and the completed form posted as soon as possible after the 0900 GMT observation on the Sunday. The address to which it should be sent is printed on the back of the form.

II.2. AUXILIARY REPORTING STATIONS

In addition to the routine detailed in 1.2 (page 6), certain selected auxiliary stations also send climatological returns to the Meteorological Office.

II.2.1. Registers of observations at auxiliary reporting stations. Depending on the type of observation made by auxiliary reporting stations for synoptic purposes, one of two Registers is used for recording the original observation: Metform 2611 (Register of Observations) or Metform 2050 (Daily Register).

The headings of the columns in the Registers are largely self-explanatory. Details of the relevant codes and methods of observing are given in *Abbreviated weather reports* which is issued to auxiliary stations making abbreviated reports. At other auxiliary stations, which make reports in the full code, the observations are entered in Metform 2050.

II.2.2. Monthly returns at auxiliary reporting stations (Metforms 3256B*, 3257B* and 3259A*).   Selected auxiliary stations send to the Meteorological Office completed monthly returns on Metforms 3259A and 3256B or 3257B. Instructions for completing these returns are given in the *Handbook of weather messages,* Part III.

Metform 3256B is a monthly return of daily observations at fixed hours, at 0000, 0300, . . . 2100 GMT, or for a selection of these hours as arranged.

Metform 3257B is a monthly return of observations at each of the 24 hours.

Metform 3259A is a monthly return of daily observations of maximum and minimum temperatures, amounts of precipitation and sunshine, certain other observations, and includes a record of significant phenomena and a weather diary.

II.2.3. Weekly returns at auxiliary reporting stations (Metforms 3110/A/ B).   Certain auxiliary reporting stations make weekly returns; Metform 3110 is used in England and Wales, Metform 3110A in Scotland and Metform 3110B in Northern Ireland, and they are dispatched to Meteorological Offices in Bracknell, Edinburgh and Belfast, respectively. Details in II.1.3 apply to the completion of this return except that there is a slight difference at auxiliary stations where maximum and minimum thermometers are normally read at 0900 and 2100 GMT. The daily extremes of temperature entered are the maximum during the 12-hour period ending at 2100 GMT, and the minimum during the 12-hour period ending at 0900 GMT on the day in question. However, the entry at the foot of the form for the highest temperature of the week refers to the 24-hour period 0900 to 0900 GMT and is not necessarily the highest of the daily maxima for the period 0900 to 2100 GMT. When the highest temperature of the week occurs in the period from 2100 to 0900 GMT it is credited to the date of commencement of the period. Similarly, when the lowest temperature of the week occurs in the period 0900 to 2100 GMT it is credited to the day on which it occurs.

II.3. METEOROLOGICAL OFFICE STATIONS

Climatological returns are completed by synoptic stations in accordance with instructions which are issued from time to time. Instructions for completing these returns are given in the *Handbook of weather messages,* Part III.

II.3.1. Daily Register (Metform 2050).   Synoptic observations are entered in the Daily Register; instructions are given in the *Handbook of weather messages,* Part III, and further useful advice is given in the front pages of the Register.

---

*See Note on page 203.

II.3.2. Monthly returns at synoptic stations (Metforms 3256B,* 3257B* and 3259A*). Selected Meteorological Office stations send to Headquarters monthly returns on Metform 3256B or 3257B, and 3259A. Instructions for completing these returns are given in the *Handbook of weather messages*, Part III.

II.3.3. Weekly returns (Metforms 3110/A/B). Selected stations make weekly returns as noted in II.2.3.

II.3.4. Visibility and cloud height summaries (Metform 2326B*). Selected stations on aerodromes make these monthly returns.

II.4. ADDITIONAL RETURNS

Some additional returns are made by stations equipped with instruments which are not widely distributed. Other returns are required from stations with a standard instrument but from which a more detailed extraction of the data is required for climatological purposes. The elements involved in these additional returns are wind direction and speed, sunshine and rainfall.

II.4.1. Analysis of anemograms (Metform 6910). This return should be made by stations with anemographs and provides an analysis of the hourly mean wind and maximum gust for each hour of the day, together with space for supplementary notes to be entered as necessary. It is completed on a monthly basis.

At some stations, where other commitments preclude this analysis, the anemogram is forwarded elsewhere for the return to be completed.

II.4.2. Tabulation of hourly precipitation rate (Metform 7113). A return of the hourly values of precipitation may be made once a month by those stations with a rain recorder—usually one of the siphoning variety. Instructions for the completion of the form are issued separately to the stations concerned and are not printed on the form itself.

II.4.3. Duration and rate of rainfall (Metform 3441). This return is made only by those stations with a siphoning rain recorder. It is usually completed once a month in conjunction with Metform 7113. This return gives information on the duration of any falls of rain exceeding certain specified amounts. Instructions for its completion are detailed on the form itself.

II.4.4. Hourly values of sunshine (Metform 3445). An analysis of the duration of sunshine for each hour of the day is made on Metform 3445. This return is completed monthly. Notes on the method of entering the values of sunshine duration are issued by the Climatological Services Branch.

Note. After an initial trial period it is hoped that the compilation of climatological archives by computer will make the completion of Metforms 2326B, 3256B, 3257B and 3259A unnecessary at synoptic stations, but weather diaries will still be required.

APPENDIX III

TABLES

Table I   Reduction of pressure in millibars from barometer level to another level

Table II(a)   Height (in metres) of cloud base measured by vertical searchlight

Table II(b)   Height (in feet) of cloud base measured by vertical searchlight

Table III   Value (to the nearest minute) of the 'equation of time' for each day of the year to give time of local noon (GMT)

Table IV   Conversion of degrees Celsius to Fahrenheit

Table V   Conversion of inches (0·00 to 9·99) to millimetres and tenths

Table VI   Conversion of metres to feet

Table VII   Conversion of knots to miles per hour and metres per second

In Chapter 7 it was noted that mercury barometers were being progressively phased out and therefore the practice of including a section on the corrections to mercury barometers was being discontinued in this edition. This policy eliminates several tables which were included in previous editions of this handbook. Observers who still have a requirement for correction tables applicable to mercury barometers will need to retain their 3rd editions or be able to refer to the first edition of the *Handbook of meteorological instruments,* Part I (1956).

Table I gives the correction (M) in millibars for the reduction of pressure from barometer level. These corrections allow the calculation of the pressure at aerodrome level (QFE—see 7.5 on page 105) or, in conjunction with the table in 7.6 on page 106, the altimeter pressure setting (QNH), and are applicable to aneroid barometers. The values of M are derived from the following formula:

$$M = 1000(10^m - 1)$$

where $m = \dfrac{h}{18429 \cdot 1 + 67 \cdot 53\,T + 0 \cdot 003\,h}$

and h = height of the station above sea level, in metres
T = temperature at the station level, in degrees Celsius.

Table I.   Reduction of pressure in millibars from barometer level to another level

Pressure at barometer level, 1000 mb*

Values are positive for reduction to a level below that of the barometer

Height	AIR TEMPERATURE (°C) (Dry bulb in screen)											
	−15	−10	−5	0	5	10	15	20	25	30	35	40
metres					*millibars*							
5	0·7	0·6	0·6	0·6	0·6	0·6	0·6	0·6	0·6	0·6	0·6	0·5
10	1·3	1·3	1·3	1·3	1·2	1·2	1·2	1·2	1·1	1·1	1·1	1·1
15	2·0	1·9	1·9	1·9	1·8	1·8	1·8	1·7	1·7	1·7	1·7	1·6
20	2·6	2·6	2·5	2·5	2·5	2·4	2·4	2·3	2·3	2·3	2·2	2·2
25	3·3	3·2	3·2	3·1	3·1	3·0	3·0	2·9	2·9	2·8	2·8	2·7
30	4·0	3·9	3·8	3·8	3·7	3·6	3·6	3·5	3·4	3·4	3·3	3·3
35	4·6	4·5	4·5	4·4	4·3	4·2	4·2	4·1	4·0	3·9	3·9	3·8
40	5·3	5·2	5·1	5·0	4·9	4·8	4·7	4·7	4·6	4·5	4·4	4·4
45	6·0	5·9	5·7	5·6	5·5	5·4	5·3	5·3	5·2	5·1	5·0	4·9
50	6·6	6·5	6·4	6·3	6·2	6·0	5·9	5·8	5·7	5·6	5·6	5·5
55	7·3	7·2	7·0	6·9	6·8	6·7	6·5	6·4	6·3	6·2	6·1	6·0
60	8·0	7·8	7·7	7·5	7·4	7·3	7·1	7·0	6·9	6·8	6·7	6·6
65	8·6	8·5	8·3	8·2	8·0	7·9	7·7	7·6	7·5	7·3	7·2	7·1
70	9·3	9·1	8·9	8·8	8·6	8·5	8·3	8·2	8·0	7·9	7·8	7·7
75	10·0	9·8	9·6	9·4	9·2	9·1	8·9	8·8	8·6	8·5	8·3	8·2
80	10·6	10·4	10·2	10·0	9·9	9·7	9·5	9·4	9·2	9·0	8·9	8·8
85	11·3	11·1	10·9	10·7	10·5	10·3	10·1	9·9	9·8	9·6	9·5	9·3
90	12·0	11·7	11·5	11·3	11·1	10·9	10·7	10·5	10·4	10·2	10·0	9·9
95	12·6	12·4	12·2	11·9	11·7	11·5	11·3	11·1	10·9	10·8	10·6	10·4
100	13·3	13·1	12·8	12·6	12·3	12·1	11·9	11·7	11·5	11·3	11·1	11·0
105	14·0	13·7	13·5	13·2	13·0	12·7	12·5	12·3	12·1	11·9	11·7	11·5
110	14·6	14·4	14·1	13·8	13·6	13·3	13·1	12·9	12·7	12·5	12·3	12·1
115	15·3	15·0	14·7	14·5	14·2	14·0	13·7	13·5	13·2	13·0	12·8	12·6
120	16·0	15·7	15·4	15·1	14·8	14·6	14·3	14·1	13·8	13·6	13·4	13·2
125	16·7	16·3	16·0	15·7	15·5	15·2	14·9	14·7	14·4	14·2	13·9	13·7
130	17·3	17·0	16·7	16·4	16·1	15·8	15·5	15·2	15·0	14·7	14·5	14·3
135	18·0	17·7	17·3	17·0	16·7	16·4	16·1	15·8	15·6	15·3	15·1	14·8
140	18·7	18·3	18·0	17·6	17·3	17·0	16·7	16·4	16·2	15·9	15·6	15·4
145	19·4	19·0	18·6	18·3	17·9	17·6	17·3	17·0	16·7	16·5	16·2	15·9
150	20·0	19·6	19·3	18·9	18·6	18·2	17·9	17·6	17·3	17·0	16·7	16·5
155	20·7	20·3	19·9	19·6	19·2	18·9	18·5	18·2	17·9	17·6	17·3	17·0
160	21·4	21·0	20·6	20·2	19·8	19·5	19·1	18·8	18·5	18·2	17·9	17·6
165	22·1	21·6	21·2	20·8	20·5	20·1	19·7	19·4	19·1	18·7	18·4	18·1
170	22·7	22·3	21·9	21·5	21·1	20·7	20·3	20·0	19·6	19·3	19·0	18·7
175	23·4	23·0	22·5	22·1	21·7	21·3	20·9	20·6	20·2	19·9	19·6	19·3
180	24·1	23·6	23·2	22·7	22·3	21·9	21·5	21·2	20·8	20·5	20·1	19·8
185	24·8	24·3	23·8	23·4	23·0	22·5	22·2	21·8	21·4	21·0	20·7	20·4
190	·25·4	24·9	24·5	24·0	23·6	23·2	22·8	22·4	22·0	21·6	21·3	20·9
195	26·1	25·6	25·1	24·7	24·2	23·8	23·4	23·0	22·6	22·2	21·8	21·5
200	26·8	26·3	25·8	25·3	24·8	24·4	24·0	23·6	23·2	22·8	22·4	22·0
205	27·5	26·9	26·4	25·9	25·5	25·0	24·6	24·2	23·7	23·3	23·0	23·6
210	28·2	27·6	27·1	26·6	26·1	25·6	25·2	24·7	24·3	23·9	23·5	23·1
215	28·8	28·3	27·7	27·2	26·7	26·3	25·8	25·3	24·9	24·5	24·1	23·7
220	29·5	28·9	28·4	27·9	27·4	26·9	26·4	25·9	25·5	25·1	24·7	24·3
225	30·2	29·6	29·0	28·5	28·0	27·5	27·0	26·5	26·1	25·7	25·2	24·8
230	30·9	30·3	29·7	29·2	28·6	28·1	27·6	27·1	26·7	26·2	25·8	25·4
235	31·6	30·9	30·4	29·8	29·3	28·7	28·2	27·7	27·3	26·8	26·4	25·9
240	32·2	31·6	31·0	30·4	29·9	29·3	28·8	28·3	27·8	27·4	26·9	26·5
245	32·9	32·3	31·7	31·1	30·5	30·0	29·4	28·9	28·4	28·0	27·5	27·1
250	33·6	33·0	32·3	31·7	31·1	30·6	30·0	29·5	29·0	28·5	28·1	27·6

*For other pressures the corrections to be applied are in proportion.

Table II(a). Height (in metres) of cloud base measured by vertical searchlight

L = 300 metres

E	300 tan E	E	300 tan E	E	300 tan E	E	300 tan E
deg	m	deg	m	deg	m	deg	m
		25	140	50	358	74	1046
1	5	26	146	51	370	74·5	1082
2	10	27	153	52	384	75	1120
3	16	28	160	53	398	75·5	1160
4	21	29	166	54	413	76	1203
5	26	30	173	55	428	76·5	1250
6	32	31	180	56	445	77	1299
7	37	32	187	57	462	77·5	1353
8	42	33	195	58	480	78	1411
9	48	34	202	59	499	78·5	1475
10	53	35	210	60	520	79	1543
11	58	36	218	61	541	79·5	1619
12	64	37	226	62	564	80	1701
13	69	38	234	63	589	80·5	1793
14	75	39	243	64	615	81	1894
15	80	40	252	65	643	81·5	2007
16	86	41	261	66	674	82	2135
17	92	42	270	67	707	82·5	2279
18	97	43	280	68	743	83	2443
19	103	44	290	69	782	83·5	2633
20	109	45	300	70	824	84	2854
21	115	46	311	71	871	84·5	3116
22	121	47	322	72	923	85	3429
23	127	48	333	73	981	85·5	3812
24	134	49	345	73·5	1013	86	4290

The tabulated values show the height in metres of the cloud base for given values of E, the angle of elevation of the spot of light observed from a point 300 m away.

A table for any other length of base-line L metres may be constructed by multiplying the values in the table by $L/300$.

Table II(b).   Height (in feet) of cloud base measured by vertical searchlight

$L = 1000$ feet

E	1000 tan E	E	1000 tan E	E	100 tan E	E	1000 tan E
deg	*ft*	*deg*	*ft*	*deg*	*ft*	*deg*	*ft*
		25	466	50	1192	74	3487
1	17	26	488	51	1235	74·5	3606
2	35	27	510	52	1280	75	3732
3	52	28	532	53	1327	75·5	3867
4	70	29	554	54	1376	76	4011
5	87	30	577	55	1428	76·5	4165
6	105	31	601	56	1483	77	4331
7	123	32	625	57	1540	77·5	4511
8	141	33	649	58	1600	78	4705
9	158	34	675	59	1664	78·5	4915
10	176	35	700	60	1732	79	5145
11	194	36	727	61	1804	79·5	5396
12	213	37	754	62	1881	80	5671
13	231	38	781	63	1963	80·5	5976
14	249	39	810	64	2050	81	6314
15	268	40	839	65	2145	81·5	6691
16	287	41	869	66	2246	82	7115
17	306	42	900	67	2356	82·5	7596
18	325	43	932	68	2475	83	8144
19	344	44	966	69	2605	83·5	8777
20	364	45	1000	70	2747	84	9514
21	384	46	1036	71	2904	84·5	10385
22	404	47	1072	72	3078	85	11430
23	424	48	1111	73	3271	85·5	12706
24	445	49	1150	73·5	3376	86	14301

The tabulated values show the height in feet of the cloud base for given values of E, the angle of elevation of the spot of light observed from a point 1000 feet away.

A table for any other length of base-line L feet may be constructed by multiplying the values in the table by $L/1000$.

Table III. Value (to the nearest minute) of the 'equation of time' for each day of the year to give time of local noon (GMT)

The value of the 'equation of time' is used in the relationship:

Time (GMT) of local noon = 1200 − 'equation of time' ± longitude correction of four minutes per degree.

Longitude correction is plus for every degree west of Greenwich and minus for every degree east of Greenwich; for example, for 18 March at a location 3°W:

Time (GMT) of local noon = 1200 + 8 + 12 = 1220 hours.

Date	Jan.	Feb.	Mar.	Apr.	May	June	July	Aug.	Sept.	Oct.	Nov.	Dec.
1	−3	−14	−13	−4	+3	+2	−4	−6	0	+10	+16	+11
2	−4	−14	−12	−4	+3	+2	−4	−6	0	+10	+16	+11
3	−4	−14	−12	−4	+3	+2	−4	−6	0	+11	+16	+11
4	−5	−14	−12	−3	+3	+2	−4	−6	+1	+11	+16	+10
5	−5	−14	−12	−3	+3	+2	−4	−6	+1	+11	+16	+10
6	−6	−14	−12	−3	+3	+1	−4	−6	+1	+12	+16	+9
7	−6	−14	−11	−2	+3	+1	−5	−6	+2	+12	+16	+9
8	−6	−14	−11	−2	+3	+1	−5	−6	+2	+12	+16	+8
9	−7	−14	−11	−2	+4	+1	−5	−6	+2	+12	+16	+8
10	−7	−14	−11	−2	+4	+1	−5	−5	+3	+13	+16	+8
11	−8	−14	−10	−1	+4	+1	−5	−5	+3	+13	+16	+7
12	−8	−14	−10	−1	+4	0	−5	−5	+3	+13	+16	+7
13	−8	−14	−10	−1	+4	0	−6	−5	+4	+14	+16	+6
14	−9	−14	−10	0	+4	0	−6	−5	+4	+14	+16	+6
15	−9	−14	−9	0	+4	0	−6	−5	+4	+14	+15	+5
16	−9	−14	−9	0	+4	0	−6	−4	+4	+14	+15	+5
17	−10	−14	−9	0	+4	−1	−6	−4	+5	+14	+15	+4
18	−10	−14	−8	+1	+4	−1	−6	−4	+5	+15	+15	+4
19	−11	−14	−8	+1	+4	−1	−6	−4	+6	+15	+15	+3
20	−11	−14	−8	+1	+4	−1	−6	−4	+6	+15	+14	+3
21	−11	−14	−8	+1	+4	−1	−6	−3	+7	+15	+14	+2
22	−11	−14	−7	+1	+3	−2	−6	−3	+7	+15	+14	+2
23	−12	−13	−7	+2	+3	−2	−6	−3	+7	+15	+14	+1
24	−12	−13	−6	+2	+3	−2	−6	−3	+8	+16	+14	+1
25	−12	−13	−6	+2	+3	−2	−6	−2	+8	+16	+13	0
26	−12	−13	−6	+2	+3	−3	−6	−2	+8	+16	+13	0
27	−13	−13	−6	+2	+3	−3	−6	−2	+9	+16	+13	−1
28	−13	−13	−5	+3	+3	−3	−6	−2	+9	+16	+12	−1
29	−13		−5	+3	+3	−3	−6	−1	+9	+16	+12	−2
30	−13		−5	+3	+3	−3	−6	−1	+10	+16	+12	−2
31	−13		−4		+3		−6	−1		+16		−3

Table IV. Conversion of degrees Celsius to Fahrenheit

°C	·0	·2	·5	·8	°C	·0	·2	·5	·8
	degrees Fahrenheit					*degrees Fahrenheit*			
40	104	104	105	105	0	32	32	33	33
					− 0	32	32	31	31
39	102	103	103	104	− 1	30	30	29	29
38	100	101	101	102	− 2	28	28	27	27
37	99	99	99	100	− 3	27	26	26	25
36	97	97	98	98	− 4	25	24	24	23
35	95	95	96	96	− 5	23	23	22	22
34	93	94	94	95	− 6	21	21	20	20
33	91	92	92	93	− 7	19	19	19	18
32	90	90	91	91	− 8	18	17	17	16
31	88	88	89	89	− 9	16	15	15	14
30	86	86	87	87	−10	14	14	13	13
29	84	85	85	86	−11	12	12	11	11
28	82	83	83	84	−12	10	10	9	9
27	81	81	81	82	−13	9	8	8	7
26	79	79	80	80	−14	7	6	6	5
25	77	77	78	78	−15	5	5	4	4
24	75	75	76	77	−16	3	3	2	2
23	73	74	74	75	−17	1	1	1	0
22	72	72	73	73	−18	0	− 1	− 1	− 2
21	70	70	71	71	−19	− 2	− 3	− 3	− 4
20	68	68	69	69	−20	− 4	− 4	− 5	− 5
19	66	67	67	68	−21	− 6	− 6	− 7	− 7
18	64	65	65	66	−22	− 8	− 8	− 9	− 9
17	63	63	63	64	−23	− 9	−10	−10	−11
16	61	61	62	62	−24	−11	−12	−12	−13
15	59	59	60	60	−25	−13	−13	−14	−14
14	57	58	58	59	−26	−15	−15	−16	−16
13	55	56	56	57	−27	−17	−17	−17	−18
12	54	54	55	55	−28	−18	−19	−19	−20
11	52	52	53	53	−29	−20	−21	−21	−22
10	50	50	51	51	−30	−22	−22	−23	−23
9	48	49	49	50	−31	−24	−24	−25	−25
8	46	47	47	48	−32	−26	−26	−27	−27
7	45	45	45	46	−33	−27	−28	−28	−29
6	43	43	44	44	−34	−29	−30	−30	−31
5	41	41	42	42	−35	−31	−31	−32	−32
4	39	40	40	41	−36	−33	−33	−34	−34
3	37	38	38	39	−37	−35	−35	−35	−36
2	36	36	37	37	−38	−36	−37	−37	−38
1	34	34	35	35	−39	−38	−39	−39	−40

Table V.   Conversion of inches (0·00 to 4·99) to millimetres and tenths

Inches	·00	·01	·02	·03	·04	·05	·06	·07	·08	·09
					millimetres					
0·0	0·0	0·3	0·5	0·8	1·0	1·3	1·5	1·8	2·0	2·3
0·1	2·5	2·8	3·0	3·3	3·6	3·8	4·1	4·3	4·6	4·8
0·2	5·1	5·3	5·6	5·8	6·1	6·4	6·6	6·9	7·1	7·4
0·3	7·6	7·9	8·1	8·4	8·6	8·9	9·1	9·4	9·7	9·9
0·4	10·2	10·4	10·7	10·9	11·2	11·4	11·7	11·9	12·2	12·4
0·5	12·7	13·0	13·2	13·5	13·7	14·0	14·2	14·5	14·7	15·0
0·6	15·2	15·5	15·7	16·0	16·3	16·5	16·8	17·0	17·3	17·5
0·7	17·8	18·0	18·3	18·5	18·8	19·1	19·3	19·6	19·8	20·1
0·8	20·3	20·6	20·8	21·1	21·3	21·6	21·8	22·1	22·4	22·6
0·9	22·9	23·1	23·4	23·6	23·9	24·1	24·4	24·6	24·9	25·1
1·0	25·4	25·7	25·9	26·2	26·4	26·7	26·9	27·2	27·4	27·7
1·1	27·9	28·2	28·4	28·7	29·0	29·2	29·5	29·7	30·0	30·2
1·2	30·5	30·7	31·0	31·2	31·5	31·8	32·0	32·3	32·5	32·8
1·3	33·0	33·3	33·5	33·8	34·0	34·3	34·5	34·8	35·1	35·3
1·4	35·6	35·8	36·1	36·3	36·6	36·8	37·1	37·3	37·6	37·8
1·5	38·1	38·4	38·6	38·9	39·1	39·4	39·6	39·9	40·1	40·4
1·6	40·6	40·9	41·1	41·4	41·7	41·9	42·2	42·4	42·7	42·9
1·7	43·2	43·4	43·7	43·9	44·2	44·5	44·7	45·0	45·2	45·5
1·8	45·7	46·0	46·2	46·5	46·7	47·0	47·2	47·5	47·8	48·0
1·9	48·3	48·5	48·8	49·0	49·3	49·5	49·8	50·0	50·3	50·5
2·0	50·8	51·1	51·3	51·6	51·8	52·1	52·3	52·6	52·8	53·1
2·1	53·3	53·6	53·8	54·1	54·4	54·6	54·9	55·1	55·4	55·6
2·2	55·9	56·1	56·4	56·6	56·9	57·2	57·4	57·7	57·9	58·2
2·3	58·4	58·7	58·9	59·2	59·4	59·7	59·9	60·2	60·5	60·7
2·4	61·0	61·2	61·5	61·7	62·0	62·2	62·5	62·7	63·0	63·2
2·5	63·5	63·8	64·0	64·3	64·5	64·8	65·0	65·3	65·5	65·8
2·6	66·0	66·3	66·5	66·8	67·1	67·3	67·6	67·8	68·1	68·3
2·7	68·6	68·8	69·1	69·3	69·6	69·9	70·1	70·4	70·6	70·9
2·8	71·1	71·4	71·6	71·9	72·1	72·4	72·6	72·9	73·2	73·4
2·9	73·7	73·9	74·2	74·4	74·7	74·9	75·2	75·4	75·7	75·9
3·0	76·2	76·5	76·7	77·0	77·2	77·5	77·7	78·0	78·2	78·5
3·1	78·7	79·0	79·2	79·5	79·8	80·0	80·3	80·5	80·8	81·0
3·2	81·3	81·5	81·8	82·0	82·3	82·6	82·8	83·1	83·3	83·6
3·3	83·8	84·1	84·3	84·6	84·8	85·1	85·3	85·6	85·9	86·1
3·4	86·4	86·6	86·9	87·1	87·4	87·6	87·9	88·1	88·4	88·6
3·5	88·9	89·2	89·4	89·7	89·9	90·2	90·4	90·7	90·9	91·2
3·6	91·4	91·7	91·9	92·2	92·5	92·7	93·0	93·2	93·5	93·7
3·7	94·0	94·2	94·5	94·7	95·0	95·3	95·5	95·8	96·0	96·3
3·8	96·5	96·8	97·0	97·3	97·5	97·8	98·0	98·3	98·6	98·8
3·9	99·1	99·3	99·6	99·8	100·1	100·3	100·6	100·8	101·1	101·3
4·0	101·6	101·9	102·1	102·4	102·6	102·9	103·1	103·4	103·6	103·9
4·1	104·1	104·4	104·6	104·9	105·2	105·4	105·7	105·9	106·2	106·4
4·2	106·7	106·9	107·2	107·4	107·7	108·0	108·2	108·5	108·7	109·0
4·3	109·2	109·5	109·7	110·0	110·2	110·5	110·7	111·0	111·3	111·5
4·4	111·8	112·0	112·3	112·5	112·8	113·0	113·3	113·5	113·8	114·0
4·5	114·3	114·6	114·8	115·1	115·3	115·6	115·8	116·1	116·3	116·6
4·6	116·8	117·1	117·3	117·6	117·9	118·1	118·4	118·6	118·9	119·1
4·7	119·4	119·6	119·9	120·1	120·4	120·7	120·9	121·2	121·4	121·7
4·8	121·9	122·2	122·4	122·7	122·9	123·2	123·4	123·7	124·0	124·2
4·9	124·5	124·7	125·0	125·2	125·5	125·7	126·0	126·2	126·5	126·7

Table V (*continued*).   Conversion of inches (5·00 to 9·99) to millimetres and tenths

Inches	·00	·01	·02	·03	·04	·05	·06	·07	·08	·09
					millimetres					
5·0	127·0	127·3	127·5	127·8	128·0	128·3	128·5	128·8	129·0	129·3
5·1	129·5	129·8	130·1	130·3	130·6	130·8	131·1	131·3	131·6	131·8
5·2	132·0	132·3	132·6	132·8	133·1	133·3	133·6	133·9	134·1	134·4
5·3	134·6	134·9	135·1	135·4	135·6	135·9	136·1	136·4	136·7	136·9
5·4	137·2	137·4	137·7	137·9	138·2	138·4	138·7	138·9	139·2	139·5
5·5	139·7	139·9	140·2	140·5	140·7	141·0	141·2	141·5	141·7	142·0
5·6	142·2	142·5	142·7	143·0	143·3	143·5	143·8	144·0	144·3	144·5
5·7	144·8	145·0	145·3	145·5	145·8	146·1	146·3	146·6	146·8	147·1
5·8	147·3	147·6	147·8	148·1	148·3	148·6	148·8	149·1	149·3	149·6
5·9	149·9	150·1	150·4	150·6	150·9	151·1	151·4	151·6	151·9	152·1
6·0	152·4	152·7	152·9	153·2	153·4	153·7	153·9	154·2	154·4	154·7
6·1	154·9	155·2	155·5	155·7	156·0	156·2	156·5	156·7	157·0	157·2
6·2	157·4	157·7	158·0	158·2	158·5	158·7	159·0	159·3	159·5	159·8
6·3	160·0	160·3	160·5	160·8	161·0	161·3	161·5	161·8	162·1	162·3
6·4	162·6	162·8	163·1	163·3	163·6	163·8	164·1	164·3	164·6	164·9
6·5	165·1	165·3	165·6	165·9	166·1	166·4	166·6	166·9	167·1	167·4
6·6	167·6	167·9	168·1	168·4	168·7	168·9	169·2	169·4	169·7	169·9
6·7	170·2	170·4	170·7	170·9	171·2	171·5	171·7	172·0	172·2	172·5
6·8	172·7	173·0	173·3	173·5	173·7	174·0	174·2	174·5	174·7	175·0
6·9	175·3	175·5	175·8	176·0	176·3	176·5	176·8	177·0	177·3	177·5
7·0	177·8	178·1	178·3	178·6	178·8	179·1	179·3	179·6	179·8	180·1
7·1	180·3	180·6	180·9	181·1	181·4	181·6	181·9	182·1	182·4	182·6
7·2	182·9	183·1	183·4	183·6	183·9	184·1	184·4	184·7	184·9	185·2
7·3	185·4	185·7	185·9	186·2	186·4	186·7	186·9	187·2	187·5	187·7
7·4	188·0	188·2	188·5	188·7	189·0	189·2	189·5	189·7	190·0	190·3
7·5	190·5	190·7	191·0	191·3	191·5	191·8	192·0	192·3	192·5	192·8
7·6	193·0	193·3	193·5	193·8	194·1	194·3	194·6	194·8	195·1	195·3
7·7	195·6	195·8	196·1	196·3	196·6	196·9	197·1	197·4	197·6	197·9
7·8	198·1	198·4	198·6	198·9	199·1	199·4	199·6	199·9	200·1	200·4
7·9	200·7	200·9	201·2	201·4	201·7	201·9	202·2	202·4	202·7	202·9
8·0	203·2	203·5	203·7	204·0	204·2	204·5	204·7	205·0	205·2	205·5
8·1	205·7	206·0	206·3	206·5	206·8	207·0	207·3	207·5	207·8	208·0
8·2	208·3	208·5	208·8	209·0	209·3	209·5	209·8	210·1	210·3	210·6
8·3	210·8	211·1	211·3	211·6	211·8	212·1	212·3	212·6	212·9	213·1
8·4	213·3	213·6	213·9	214·1	214·4	214·6	214·9	215·1	215·4	215·7
8·5	215·9	216·1	216·4	216·7	216·9	217·2	217·4	217·7	217·9	218·2
8·6	218·4	218·7	218·9	219·2	219·5	219·7	220·0	220·2	220·5	220·7
8·7	221·0	221·2	221·5	221·7	222·0	222·3	222·5	222·8	223·0	223·3
8·8	223·5	223·8	224·0	224·3	224·5	224·8	225·0	225·3	225·5	225·8
8·9	226·1	226·3	226·6	226·8	227·1	227·3	227·6	227·8	228·1	228·3
9·0	228·6	228·9	229·1	229·4	229·6	229·9	230·1	230·4	230·6	230·9
9·1	231·1	231·4	231·7	231·9	232·2	232·4	232·7	232·9	233·2	233·4
9·2	233·7	233·9	234·2	234·4	234·7	234·9	235·2	235·5	235·7	236·0
9·3	236·2	236·5	236·7	237·0	237·2	237·5	237·7	238·0	238·3	238·5
9·4	238·8	239·0	239·3	239·5	239·8	240·0	240·3	240·5	240·8	241·1
9·5	241·3	241·5	241·8	242·1	242·3	242·6	242·8	243·1	243·3	243·6
9·6	243·8	244·1	244·3	244·6	244·9	245·1	245·4	245·6	245·9	246·1
9·7	246·4	246·6	246·9	247·1	247·4	247·7	247·9	248·2	248·4	248·7
9·8	248·9	249·2	249·4	249·7	249·9	250·2	250·4	250·7	250·9	251·2
9·9	251·5	251·7	252·0	252·2	252·5	252·7	253·0	253·2	253·5	253·7

For values between 10 and 20 inches add 254·0 to the numbers above. Thus 15·32 inches (10 + 5·32) = 254·0 + 135·1 = 389·1 mm.

For amounts between 20 and 30 inches add 508·0; between 30 and 40 inches add 762·0; between 40 and 50 inches add 1016·0 to the numbers above.

Table VI.   Conversion of metres to feet

1 metre = 3·280 839 9 feet = 39·370 079 inches

Metres	Feet	Metres	Feet	Metres	Feet	Metres	Feet
1	3·3	10	32·8	100	328·1	1 000	3 280·8
2	6·6	20	65·6	200	656·2	2 000	6 561·7
3	9·8	30	98·4	300	984·3	3 000	9 842·5
4	13·1	40	131·2	400	1312·3	4 000	13 123·4
5	16·4	50	164·0	500	1640·4	5 000	16 404·2
6	19·7	60	196·9	600	1968·5	6 000	19 685·0
7	23·0	70	229·7	700	2296·6	7 000	22 965·9
8	26·2	80	262·5	800	2624·7	8 000	26 246·7
9	29·5	90	295·3	900	2952·8	9 000	29 527·6
						10 000	32 808·4

m	0·1	0·2	0·3	0·4	0·5	0·6	0·7	0·8	0·9
ft	0·33	0·66	0·98	1·31	1·64	1·97	2·30	2·62	2·95

Table VII.   Conversion of knots* to miles per hour and metres per second

1 knot = 1·15078 miles per hour = 0·51444 metres per second

Knots	0	1	2	3	4	5	6	7	8	9
				miles per hour						
0	0·0	1·2	2·3	3·5	4·6	5·8	6·9	8·1	9·2	10·4
10	11·5	12·7	13·8	15·0	16·1	17·3	18·4	19·6	20·7	21·9
20	23·0	24·2	25·3	26·5	27·6	28·8	29·9	31·1	32·2	33·4
30	34·5	35·7	36·8	38·0	39·1	40·3	41·4	42·6	43·7	44·9
40	46·0	47·2	48·3	49·5	50·6	51·8	52·9	54·1	55·2	56·4
50	57·5	58·7	59·8	61·0	62·1	63·3	64·4	65·6	66·7	67·9
60	69·0	70·2	71·3	72·5	73·6	74·8	76·0	77·1	78·3	79·4
70	80·6	81·7	82·9	84·0	85·2	86·3	87·5	88·6	89·8	90·9
80	92·1	93·2	94·4	95·6	96·7	97·8	99·0	100·1	101·3	102·4
90	103·6	104·7	105·9	107·0	108·2	109·3	110·5	111·6	112·8	113·9
100	115·1	116·2	117·4	118·6	119·7	120·8	122·0	123·1	124·3	125·4
				metres per second						
0	0·0	0·5	1·0	1·5	2·1	2·6	3·1	3·6	4·1	4·6
10	5·1	5·7	6·2	6·7	7·2	7·7	8·2	8·7	9·3	9·8
20	10·3	10·8	11·3	11·8	12·3	12·9	13·4	13·9	14·4	14·9
30	15·4	15·9	16·5	17·0	17·5	18·0	18·5	19·0	19·6	20·1
40	20·6	21·1	21·6	22·1	22·6	23·1	23·7	24·2	24·7	25·2
50	25·7	26·2	26·8	27·3	27·8	28·3	28·8	29·3	29·8	30·3
60	30·9	31·4	31·9	32·4	32·9	33·4	34·0	34·5	35·0	35·5
70	36·0	36·5	37·0	37·6	38·1	38·6	39·1	39·6	40·1	40·6
80	41·2	41·7	42·2	42·7	43·2	43·7	44·2	44·8	45·3	45·8
90	46·3	46·8	47·3	47·8	48·4	48·9	49·4	49·9	50·4	50·9
100	51·4	52·0	52·5	53·0	53·5	54·0	54·5	55·0	55·6	56·1

*The United Kingdom in 1970 adopted the International Nautical Mile of 1852 metres in place of the Admiralty Mile of 6080 feet (1853·18 metres); this table therefore differs slightly from that in previous editions of the handbook.

BIBLIOGRAPHY

London, Meteorological Office. Handbook of meteorological instruments, Part I: Instruments for surface observations. London, HMSO, 1956 (reprinted 1962). (See also 2nd edition below.)

London, Meteorological Office. Hygrometric tables, Part II: For use with Stevenson screen readings in degrees Celsius, 2nd edition. London, HMSO, 1964 (reprinted 1978).

London, Meterorological Office. Hygrometric tables, Part III: For aspirated psychrometer readings in degrees Celsius, 2nd edition. London, HMSO, 1964 (reprinted 1978).

London, Meteorological Office. The measurement of upper winds by means of pilot balloons, 4th edition. London, HMSO, 1968.

London, Meteorological Office. Abbreviated weather reports. London, HMSO, 1982.

London, Meteorological Office. Handbook of weather messages, Part II: Codes and specifications, 7th edition. London, HMSO, 1979. (Amendment Lists are issued from time to time.)

London, Meteorological Office. Handbook of weather messages, Part III: Coding, 6th edition. London, HMSO, 1979. (Amendment Lists are issued from time to time.)

London, Meteorological Office. Handbook of meteorological instruments, 2nd edition (published in eight separate volumes). Vol. 1: Measurement of atmospheric pressure; Vol. 2. Measurement of temperature; Vol. 3: Measurement of humidity; Vol. 4: Measurement of surface wind; Vol. 5: Measurement of precipitation and evaporation; Vol. 6: Measurement of sunshine and solar and terrestrial radiation; Vol. 7: Measurement of visibility and cloud height; Vol. 8: General observational systems. London, HMSO, 1980–82.

London, Meteorological Office. Cloud types for observers, Revised edition (colour). London, HMSO, 1982.

London, Meteorological Office. Dew-point tables for screen readings, degrees Celsius. 1982.

London, Meteorological Office. Met. O. Leaflet No. 11: The Meteorological Office Calendar (yearly).

London, Nautical Almanac Office. The nautical almanac. London, HMSO, yearly.

Washington, D.C., Smithsonian Institution. Smithsonian meteorological tables, 6th revised edition. Washington, D.C., 1958.

Geneva, World Meteorological Organization. Guide to meteorological instrument and observing practices, 4th edition. WMO – No. 8, 1971.

Geneva, World Meteorological Organization. International cloud atlas, Vol. I: Manual on the observation of clouds and other meteors, Revised edition. WMO – No. 407, 1975.

INDEX

Page

Abbreviated weather reports 6
Aerodrome level 105
Afterglow 166
Air discharge (streak lightning) 66
Airfield, see Aerodrome
Alidade 33, 37
Alpine glow 166
Altimeter setting (QNH) 106
Altitude of station 178–180
Altocumulus 22
Altostratus 23
Anemograms 4, 17, 203
Anemographs (see also Anemometer) 91–94, 197

Anemometer
  cup 88, 89, 90
  effective height 82, 197
  hand 89
  reading from dials 90
  siting 82, 178, 197
Anti-condensation shield 126
Arcs
  auroral 171, 172
  circumzenithal 163
  of contact 161, 162
  paranthelic 161, 162
Atmospheric
  obscurity 78
  pressure, see Pressure
Aurora 170–173
Autographic
  instruments 4–5, 15–17, 129
  records 17

Balloon
  cloud height measured by 11, 31, 37
  filling shed 39
  inflation of 40
  observation of upper winds by 5
  rates of ascent 37, 41
  storage of 41
Bare ground (plot) 95, 177
Barograms 5, 17, 109
Barograph 106–109, 184
Barometer
  aneroid 99–102
    certification 183
    schematic drawing 100
    siting and installation of 102, 184
  checking procedures 105
  mercury
    certification 183
    corrections to 103

Page

Kew-pattern and Fortin 103
Beaufort letters 72–79
  for climatological records 57, 200
  in Health Resort reports 8, 72
  in Register 57
  in Press report at 6 p.m. 79
Beaufort scale of wind speeds 86–87
Bishop's ring 167
Blizzard 71
Brocken spectre 168

Calendar, Meteorological Office 5
Candela (luminous intensity) 52
Cirrocumulus 22
Cirrostratus 22
Cirrus 21
Climatological
  records and returns 200–203
  stations 7, 57, 72, 183, 200–201
Cloud
  accessory 28
  altitude 20
  amount 29
  appearance of 19, 26
  base 31
    recorder 41
  below station 19
  classification of 18, 21
  composition of 18
  colour 20, 165, 168
  definition of 18
  discharges 66
  funnel 28, 63
  genera 21, 33
  height 14, 20, 31–33
    balloon, see Balloon
  levels 20
  luminance 19
  mother 29
  mother-of-pearl 168
  nacreous 168, 170
  noctilucent 169, 170
  observation of 14, 19, 21, 32
  photography 175–176
  precipitation of 29, 68
  searchlight 31, 33–36, 206, 207
  special 168
  species 24
  supplementary features 28
  varieties 26
  vertical extent 20
Coastal stations, see Station

214

Page

Concrete slab
  minimum thermometer 128, 188
  state of 97
Conversion tables
  °C to °F 209
  inches to mm and tenths 210–211
  knots to mile/h and m/s 212
  metres to feet 212
Coronae 22, 24, 167
Corposant (St Elmo's fire) 170
Corrections
  aneroid barometers 101
  cloud height at aerodromes 31
  hygrograph readings 121
  mercury barometers 204
  thermometers
    aspirated and inspectors' 110, 113
  wind to standard height
    land stations 83
    sea stations 84
Counterglow 166, 175
Counter-sun 162
Crepuscular rays 173
Cumulonimbus 24
Cumulus 24
  tropical 32

Day
  constant, definition of 156
  darkness 173
  first observation of 11
  length of 13
Daylight, duration of 13
Declination
  magnetic 180, 182
  of sun 190–194
Desiccator, silica-gel 37
Desynn 91
Dew 61, 96
  collected by rain-gauge 135, 140
Dew-point 119, 124, 133
Diamond dust 60
Diffraction 160
Discharge
  air, cloud and ground 66
Drizzle
  description of 59
  freezing 59
  intensity of 71, 72
  supercooled 63
Dust
  devil 65
  drifting or blowing 64
  haze 64

Page

  in suspension, diffraction by 160
  whirl 63, 65
Duststorm 65

Earth shadow 166
Effective height of anemometer 82, 197
Electrometeors 65, 75, 170–173
Evaporation, temperature of 119
Eye-level, definition of 61

Fallstreaks 28
Fata Morgana 165
Fiducial point 103
Fog
  definition of 60
  droplets, deposit of 61
    in rain-gauge 135, 140
  freezing 63
  high 173
  ice 60
  in past hour 67
  shallow 60
  wet, on thermometers 124
Frost, hoar, see Hoar frost
Funnel cloud 28, 63

Gale 81, 86–87
Gegenschein 174
Glaze 63
Glory 167
Green flash 166
Grid reference 179
Ground
  bare patch 95, 177
  discharges 66
  ice 63
  state of 95–96
Gust
  definition of 82
  on anemograph record 92
  speed uncorrected for synoptic
    purposes 83
Gustiness 92

Hail
  association with cumulonimbus .. 24, 29
  definition of 60
  reporting of 71, 72
  small 60
  stones, shapes and sizes 60
  water equivalent measured 142, 151

Page

Halo phenomena 22, 60, 160–163, 167
  photographing 175
Haze 64, 78
Health Resort Scheme 8, 200
Hectopascal 98
Helium, use of 41
Hoar frost 62, 96
  advection 62
  deposits in rain-gauge 67, 140
Hours, numbering of 11
Humidity, relative 121, 132
  determination of 119
  haze/mist criteria 78
  obtained from hygrograph 121, 130
Humidity slide-rule 119, 133
Hydrogen, use of 37–41
Hydrometeors 18, 58, 73
Hygrograms 5, 17, 131
Hygrograph, hair 108, 130, 131
  used instead of supercooled wet bulb 121
Hygrometer
  aspirated 132
  non-aspirated 119
Hygrometric tables 119, 133

Ice
  bulb thermometer 120, 121
  clear 62
  crystals 60, 160, 163
  deposit 62–63
  fog 60
  ground 63
  particles 59–60, 67
  pellets 59, 135, 142
Inspection of stations 105, 113
Instrument
  autographic 4–5, 15–17, 129
  enclosure 177, 179
Instrumental equipment 183
International
  Cloud Atlas 18, 58, 160
  Nautical Mile 212
  symbols for meteors 73
Irisation (iridescence) 168
Isogonals 180, 182

Jacob's ladder 173

Kelvin, definition of SI unit 110
Koschmieder's formula 44, 54

Lightning 66

Page

Line-squall 67
Lithometeors 63–65, 74
Looming 164
Lull 82, 92

Magnetic
  declination 180, 182
  tape event recorder 151
Magslip 91
Mean sea level, reference datum 180
Measure, rain, see Rain measure
METAR, codes and wall-card 6, 9
Meteorological Office Calendar 5
Meteorological optical range (MOR) .. 43
Meteor, definition of 58
Millibar 98
Mirage 164
Mist 61
Mock moons and suns 161, 162
Mock-sun ring 161, 162
Mother-clouds 29
Mother-of-pearl clouds 168

Nautical Almanac 13, 181
Nebule, definition of 48
Nimbostratus 23
Nomograms to determine visibility .. 53–55
North, true, determination of 181
Northern lights 171

Observation (see also Reports)
  at night 19, 20, 31, 32, 51, 114
  at stations making abbreviated reports .. 6
  avoidance of errors in 13
  climatological stations 7, 10
  entry, checking of 13
  first of the day 11
  for aviation 6, 9
  hours of 10, 11
  official record of 13
  order of making 3, 6, 7
  pilot-balloon 5, 11, 31
  recording 5, 200–203
  site 177–179
  special phenomena 175
  standard hours of 9
  time of, actual and official 11
Obstruction lights 52
Okta (unit for reporting cloud amount) .. 30
Ordnance Survey maps, use of 32, 178, 181
Orientation 180–182

Page

Parallax
  rainfall reading 139
  soil thermometer reading 128
  thermometer reading 114
Paranthelia 161, 162
Parantiselenae 162
Paraselenae 162
Paraselenic circle 162
Parhelia 161, 162
Parhelic circle 161, 162
Pascal, definition of SI unit 98
Past weather 72
Phenomena
  descriptions of 58–63, 160–175
  recording, photographing
    .... 170, 171, 175–176
  within sight 78
Photography 170, 171, 175–176
Photometeors 74–75, 160–168
Pilot balloon, see Balloon
Precipitation
  association with cloud types
    .... 23, 24, 28, 29, 68
  continuity of 68, 77
  definition of 67
  effect on balloon ascents 37
  hourly 203
  intensity of 68–72, 76
  measurement of 3, 10, 135–144
  mixed 72, 77
  observations 3, 57, 67–72, 76–78, 135
  site peculiarities affecting .. 178, 188–189
  trace of 140
  types of 58–60, 76
  within sight 68, 78
Precision aneroid barometer, see Barometer
Press report, 6 p.m. 79, 122
Pressure
  at aerodrome level (QFE) 105
  atmospheric 98
  conversion factors 98
  reduction to other levels .. 101, 105, 205
  vapour 119
Psychrometer
  aspirated 132–134
  non-aspirated 119
Purple light 165–166

QFE 105
QFF 99
QNH 106

Radiation points 27

Page

Rain
  classification of intensity 69, 72
  continuity of 77
  description of 58
  drops supercooled 63
  freezing 59
  measure
    care of 140
    housing 16
    reading 139
    types of 138, 139
  measurement after freezing .. 142, 143
  recorder, see Rain recorder
Rainbows 163
Rain-gauge
  Bradford 138
  bottle 140, 141
  care of 140–141
  catch due to surface condensation 135
  climatological 3, 137, 141
  gravimetric 152
  hours of reading 141
  measurement of solid
    precipitation 142–144, 151
  Meteorological Office preferred 136, 137
  Mk 2 and Mk 3 137
  Octapent 136, 138
  rates of accumulation in 69–70
  recording, see Rain recorder
  site and exposure 135, 178, 188
  Snowdon 136, 138
  tipping-bucket 148–151
    magnetic-tape event recorder 151
  turf wall for 189
Rain recorder, tilting-siphon 145–148
  assessment of intensity from 69–70
  frost protection 146
  sample chart traces 69
  siting and exposure 189
Rayleigh scattering 166
Red flash 167
Refraction 160
Register
  climatological station 8, 57, 200
  corrections 13
  Daily 5, 13, 201, 202
    official time entered in 11
  for abbreviated synoptic reports .. 6, 201
  Health Resort stations 8, 200
  of Observations 6
  official record of surface observations ..13
  Pocket 8, 200
  remarks column 58, 72
  use of Beaufort letters in 57, 76–78

Page

Reports (see also Observation)
abbreviated synoptic 6
Air Traffic Control 99
aviation 6
from agrometeorological stations 8
hourly rainfall 3
METAR 6, 9
simplified, plain-language 7
special at 6 p.m. 8, 10, 79
sudden changes 6
supplementary data, 0900 and 2100
GMT 4
Returns
additional 203
monthly and weekly 201, 202, 203
Rime
deposit in rain-gauges 135, 140
description of 62
Run of wind 8, 89, 197
Runway visual range 56

St Elmo's fire 170
Sand
blowing or drifting 64, 65
whirl 63, 65
Sandstorm 65
Scintillation 165
Screen, thermometer
approved types 117, 185–186
care of 118
exposure and installation 185–186
use for storage 16
Shimmer 165
Showers
definition of 68
intensity of 70, 71, 76
Site, observation 177–179
Sky
obscured 55, 76
state of 67, 76
ugly threatening 76
Sleet 57
Smog 60
Smoke 64, 173
Snow
association with cloud types 29
classification of intensity 71–72
cover, rate of increase 71
definition of 59
drifting and blowing 61
grains 59
measurement of 135, 137, 142–144
on ground 63, 67, 96
pellets 59
rainfall equivalent of 142–144

Page

Snow depth
measurement of 152
when reported by synoptic stations 3, 4, 152
Snow lying
climatological observation of 67, 142, 152
day of 67
description of 67
fresh 152
rainfall equivalent of 142, 143–144
Southern lights 171
Spout 63
Spray 61
Squall 67, 82
State of
concrete slab 97
ground 95–96
sky 76
Static pressure head 102, 184
Station
agricultural meteorological 8, 201
authority at 177
auxiliary reporting 6, 177, 201–202
climatological .. 7, 57, 72, 183, 200–201
coastal
rain-gauge exposure 178
visibility 45, 47
wind speed 88, 178
co-ordinates 178–180
Health Resort 8, 10, 200
height 178
inspection of 105, 113
layout 179
synoptic 3, 57, 72, 183, 202–203
Storm 86–87
Stratus 23
Stratocumulus 23
Sublimation of water vapour 62
Sudden or significant changes 6, 57, 90
Sun
coloration 64
in transit 156
mock 161, 162
pillar 163
Sunbeams 173
Sunrise and sunset, definitions of 13
Sunshine recorder
bowl, cross-section 155
cards 154–155
care and adjustment 158–159
siting and installation 178, 190–196
support for 195
trace at different times of year 159
types of 153
Supercooling
droplets and deposits of ice 62, 63
effect on wet bulb 121

Page

Temperature (see also Thermometer)
  by aspirated psychrometer/
    hygrometer 132
  conversion °C to °F 209
  effect of jet aircraft on 125
  indicator 115, 116
  of evaporation 119
  rounding to nearest whole degree 116
  scales 110
  site peculiarities affecting 177
  throwing to the odd 117
Theodolite, use of 175, 194
Thermogram 5, 17
Thermograph 129, 130–131
Thermometers
  accuracy 110
  attached 103
  bare-soil minimum 16, 127, 187
  bulbs above ground 117, 185
  care of 124
  category to be used 110
  certificates 110, 113
  check of 113, 116
  concrete-slab minimum 128, 188
  discrepancies in readings 124
  dry-bulb 118, 124
  electrical resistance 115–116
  estimation to tenths 114, 115
  grass minimum 112, 125–127
    anti-condensation shield 126
    check of 126
    siting 125–126, 187
    supports for 126, 127, 187
  ice bulb 120–121
  indicator 115–116
  inspectors' 110, 113
  maximum 110, 113, 122
  minimum 110, 113, 122–124
  misreading of, avoidance .. 13, 114, 124
  order of reading 124
  range of temperature scales 185
  reading of 114, 124–125, 133
  screen, see Screen
  soil 111, 112, 128–129, 186–187
  spirit, removal of bubbles 123
  wet-bulb
    aspirated 132–133
    depression 119
    water for 119, 133
    wick 119, 120, 132
Thunder 66
Thunderbolt 66
Thunderstorm
  at the station 65, 67, 76
  definition of 65

during past hour 65
intensity of 77
Time
  equation of 156
  Greenwich Mean 2, 9
  Local Apparent 155, 158, 195, 196
  Local Mean 156
  of observation 9, 10, 11
  zone 9, 11, 12
Time marks on charts 4, 17
  anemogram 5, 93
  barogram 108
  hygrogram 130, 131
  thermogram 129, 131
  tilting-siphon rain recorder 145
Tornadoes 63
Transmissometer 51
Twilight
  arch 165, 166
  astronomical 166
  colours 165
  crepuscular rays 173
  phenomena at 174
Turf wall 189
Twinkling 165

Ulloa's ring 168
Undersun 163
Units
  atmospheric pressure 98
  cloud amount 30
  height of cloud base 31
  precipitation 135
  snow depth 135, 152
  temperature 110
  visibility 44
  wind direction and speed 80, 88

Vapour pressure 119
Virga 28
Visibility
  definition of 43
  determination of 46–55
  from transmissometer reading 51
  in certain weather conditions 60, 61, 64, 71
  lights 48
  meter 48–51
  objects 44–46, 47
  varying in different directions 47
  vertical 55
Visual extinction coefficient 43
Visual threshold 52

Page

Water
  container, wet-bulb 120
  equivalent of solid precipitation .. 142–144
  purified 119
  vapour 60, 61, 62
Weather
  diary 57, 81, 200
  past 72
  present 67–72
Whirlwind 63
Wick, see Thermometers, wet-bulb
Wind
  direction
    backing or veering 82
    estimation of 84
    readings of 91, 93
    reports of 80
    sudden change in 81, 91
  force in blizzards 71

Page

Mk 5B system 94
run of 8, 89, 197
speed
  Beaufort scale 86–87
  conversions 88, 212
  estimation of 84, 85
  measurement of 80, 88–90
  reports of 80
  sudden change in 81, 91
  squall 67, 82
  standard height 80, 82, 84
  terminology 81–82
  vane 84–85, 196–197

Zodiacal
  light 174
  band and Gegenschein 174

Printed in the United Kingdom by
HMSO, Edinburgh Press

Dd 291250   C5   9/89   (271994)